Alcides Loureiro Santos

ChemSketch in Chemistry Teaching

AF294744

Alcides Loureiro Santos

ChemSketch in Chemistry Teaching

A great tool for molecular structuring

ScienciaScripts

Imprint

Any brand names and product names mentioned in this book are subject to trademark, brand or patent protection and are trademarks or registered trademarks of their respective holders. The use of brand names, product names, common names, trade names, product descriptions etc. even without a particular marking in this work is in no way to be construed to mean that such names may be regarded as unrestricted in respect of trademark and brand protection legislation and could thus be used by anyone.

Cover image: www.ingimage.com

This book is a translation from the original published under ISBN 978-3-330-75855-1.

Publisher:
Sciencia Scripts
is a trademark of
Dodo Books Indian Ocean Ltd. and OmniScriptum S.R.L publishing group

120 High Road, East Finchley, London, N2 9ED, United Kingdom
Str. Armeneasca 28/1, office 1, Chisinau MD-2012, Republic of Moldova, Europe
Printed at: see last page
ISBN: 978-620-8-30244-3

Copyright © Alcides Loureiro Santos
Copyright © 2024 Dodo Books Indian Ocean Ltd. and OmniScriptum S.R.L publishing group

CONTENTS

SUMMARY

The use of Information and Communication Technologies (ICTs) has been increasingly considered in chemistry teaching. As media tools in the teaching and learning process, they can contribute to the teacher's work and can encourage students to take part in this process in a more interactive and relevant way. Educational *software* enables certain tasks to be carried out, promoting a better understanding of school content and guaranteeing students a more meaningful and reflective education. Especially in organic chemistry, molecular structuring programs are capable of virtualizing flat and three-dimensional structures, easing the difficulty of understanding the representational level of chemistry. Among these *software programs*, *ChemSketch* is one of the most accessible and has great potential for use in schools. The research question of this work is: How can the *ACD/ChemSketch Freeware software be* used as a media tool in the process of teaching Organic Chemistry in Secondary Schools, considering its potential and limitations, in the context of Basic Education in the state of Acre? With this in mind, the main aim of this project is to contribute to the process of teaching organic chemistry through the use of *ChemSketch*. The research was carried out in four stages: in-depth study of the *ChemSketch software*; preparation of the Practical Guide to using *ChemSketch* to teach Organic Chemistry (product); dissemination and evaluation of the Practical Guide to using *ChemSketch* through a mini-course; preparation of the final version of the Practical Guide to using *ChemSketch*. It was found that most of the participating teachers did not use *software* in their classes, were not familiar with *ChemSketch* and had not taken part in continuing training courses that dealt with ICTs in chemistry teaching. However, it was argued that chemistry programs are important tools for encouraging students to learn chemistry, enabling more dynamic and interactive lessons. It also emerged that initial and continuing training courses need to address ICTs in a more practical way, encouraging the use of these tools in basic education.

Keywords: *ChemSketch*. Teaching Chemistry. ICTs.

1 INTRODUCTION

1.1 Research Presentation

The massification of information technology in modern society has provided rapid access to the knowledge built up by humanity. Even with the ease with which information can be accessed, it is important that the use of Information and Communication Technologies (ICTs) is studied and evaluated, especially in a contemporary educational environment.

In chemistry, as in other sciences, the use of ICT can contribute to teaching and student learning. There are countless possibilities. For example, *software* can be used to carry out virtual experiments, to investigate the properties of substances and to visualize molecules in three dimensions.

Molecular structure programs allow you to draw molecular structures, and some of them also allow you to manipulate and visualize molecules in three dimensions (3D). These features can be used by teachers to make it easier for students to learn certain topics in chemistry and other subjects.

One such piece of *software* is *ACD/ChemSketch Freeware*. This molecular structuring program allows you to carry out various actions that can be worked on in a school environment. *ChemSketch* can help teachers, students and researchers because, as well as allowing them to draw chemical structures, it enables them to investigate various physical and chemical properties of these substances.

This book - **Chemsketch in the Teaching of** CHEMISTRY - presents the master's research entitled "THE USE OF THE CHEMSKETCH SOFTWARE AS A TOOL IN THE TEACHING OF ORGANIC CHEMISTRY IN THE BASIC EDUCATION OF THE STATE OF ACRE". Its results could encourage elementary school chemistry teachers in particular to adopt this program as a tool within the context of teaching and learning chemistry.

1.2 Research question

The research question that this work seeks to answer is: How can the *ACD/ChemSketch Freeware software* be used as a media tool in the process of teaching Organic Chemistry in High School, considering its potential and limitations, in the context of Basic Education in the state of Acre?

1.3 Theoretical assumption

1.3.1 *New Information and Communication Technologies: potential for teaching chemistry*

1.3.1.1 Chemistry, chemistry teaching and its contributions to citizen education

Chemistry is a dynamic science from the macro to the microscopic. However, studying it is not attractive to most primary school students. Del Pino and Frison (2011) propose that this is due to the way in which chemistry has historically been taught in schools. They state that,

Traditional chemistry teaching is the result of a historical process of repeating formulas, definitions and classifications, an apparently successful didactic proposal if the purpose is to memorize information. The distribution of electrons in the atom's extra-nuclear structure, the classification of substances and chemical reactions, and chemical calculations involving the direct application of mathematical formulas are some examples of this proposal. By treating chemistry solely from a formal point of view, traditional teaching leaves out real phenomena. It is a blackboard chemistry where anything is possible (DEL PINO & FRISON, 2011, p. 2).

This shows that the teaching of chemistry has been disconnected from the reality of the students, because, approached in this content-based way, it makes it difficult for students to recognize chemistry in everyday processes and the reflection of this science in technologies and society. In this respect, Rezzadori and Cunha (2005) warn of the bad reputation of chemistry teaching, as they consider that there is a low quality and quantity of teaching proposals, which are generally restricted to the content of textbooks.

However, there is no way we can admit that, because of this situation, chemistry is losing its importance in society and its progress. Del Pino and Frison (2011) point out that the modern context in which society is inserted demands a series of new products and processes, in which chemical knowledge of matter and its transformations are essential. The great challenge is, as teachers and lovers of chemistry, to overcome the obstacles that hinder students' interest in this science.

Overcoming this will not come easily. It's not just the need for greater effort, but also for a better understanding of chemistry teaching and how it can be approached in a more attractive and dynamic way. In this sense, it is first important that the structuring concepts of chemical knowledge are clear. Mortimer and Machado (2007) present these focuses in a triangular relationship (Figure 1).

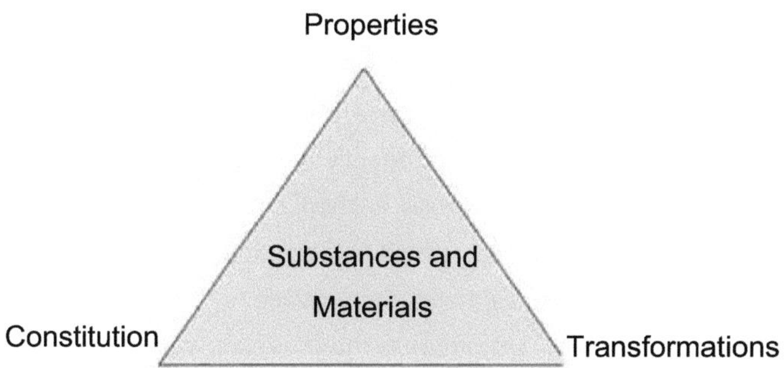

Figure 1. Foci of interest in chemistry.

Source: Mortimer and Machado (2007).

It is clear that, even with its well-defined focus, chemistry is an interdisciplinary science. The substances and materials it studies are part of a whole social context that involves economic, technological and environmental factors. Del Pino and Frison (2011) and Mortimer and Machado (2007) bring the same reflection from another perspective: the levels of Chemical Knowledge. From this perspective, chemistry is identified as a science with phenomenological, theoretical and representational characteristics (Figure 2).

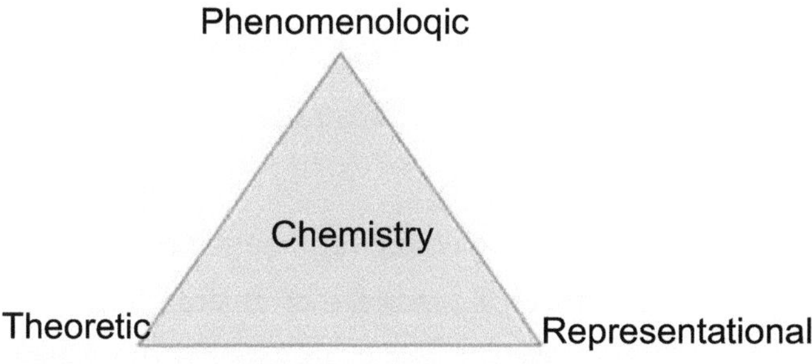

Figure 2. Aspects of chemical knowledge.

Source: Adapted from Mortimer and Machado (2007).

The importance of each of these aspects of Chemical Knowledge is presented by Del Pino and Frison (2011).

The phenomenological aspect concerns the phenomena of interest to chemistry. These are not limited to those that can be reproduced in the laboratory, but can also be materialized in social activities, such as in supermarkets, petrol stations, among others. The theoretical aspect relates to information of an atomic-molecular nature, involving explanations based on abstract models and including entities that are not directly observable such as atoms, molecules, ions, electrons, among other chemical entities. Chemical content of a symbolic nature is grouped into the representational aspect, which includes information inherent to the language of chemistry, such as chemical formulas and equations, representations of models, graphs and mathematical equations. Language is the basic knowledge that allows communication between teacher, student and educational materials through the sharing of meanings (DEL PINO & FRISON, 2011, p. 4).

Understanding and reflecting on these aspects is essential to the success of the process of teaching and learning chemistry. It loses its essence if these three levels are not properly interrelated. It is up to the teacher to homogenize the theoretical, representational and phenomenological levels. In school practice, this can produce a mixture that enables students to learn chemistry in a more meaningful and enjoyable way.

In this sense, the theoretical level explains the phenomenological by starting from the premise that there are submicroscopic entities that are interacting. The representational level has the function of putting the theoretical in writing, in formulas, drawings and figures of its own. For Bona (2009), understanding chemistry is difficult, mainly due to its high degree of abstraction, characterized by the representational level.

This situation shows the importance of working on the representational level of chemistry in a different way. The fundamental presence of abstract elements requires students to have cognitive levels that reach the formal operations suggested by Jean Piaget (WADSWORTH, 1995). Even in the abstract, educational methodologies can facilitate learning at this level of chemistry. Digital models, practical and fun activities are alternatives to the traditional form of teaching still present in Brazilian schools.

Therefore, it is necessary to use a conceptual organizational model that addresses the themes within the context of the Brazilian reality, involving interdisciplinary aspects and the Science, Technology and Society movement, CTS. This new paradigm of education for citizenship significantly alters the current form of chemistry teaching, proposing a different approach to content, methodologies, organization of the teaching-learning process and assessment methods. The methodologies used to teach chemistry should therefore not seek to merely pass on content, but aim to teach it in a way that enables citizens to be formed, allowing chemical knowledge to be present in their conceptions of themselves as unique and social human beings (SCHNETZLER, 2002).

Experimentation in chemistry, as in other areas of knowledge, allows students to better understand the phenomena they observe through their ability to abstract them. These activities aim to bring concrete objects closer to formulated theoretical conceptions, creating idealizations and, consequently, producing more knowledge about these objects (MALDANER, 2000). In this sense, it is clear that experimental practices can be a good alternative for promoting learning in a more attractive and interactive way.

In addition to experimental practice, chemistry teachers can look for other teaching methods that can make the classroom environment more productive. In this sense, Rezzadori and Cunha (2005) state:

The beginning of a change in pedagogical practice can start with the production of teaching material to be used in the classroom, as this task puts the professional in front of a set of choices that contribute greatly to their training and improve the quality of teaching. Among these choices are decisions about the type and complexity of the school content to be taught, the space/time and the resources available. By producing and experimenting with teaching materials that he or she has developed, the teacher not only assesses the quality and efficiency of the materials to be used, but also shows that he or she is a professional committed to transforming the pedagogical process at school. It is also the teacher's task to involve students in discussions about problems that are close to their hearts (REZZADORI & CUNHA, 2005, p. 178).

The advance of information technology has brought a potential tool to become part of the teaching resources available to teachers: the computer. Eichler and Pino (2000) comment on this subject:

[...] it is assumed that the combination of technological and human resources with the availability of qualified tools for learning could result in innovations in teaching and/or learning strategies and methodologies in the various areas of knowledge [...].

[...] in today's technological landscape, there are various alternatives for learning using computers, such as communicating and consulting information distributed via the Internet or using educational *software* (EICHLER & PINO, 2000, p. 835).

Still emphasizing the use of computers as a tool in the teaching and learning process, Giordan (1999) describes that:

We believe that computer simulations can be orchestrated in conjunction with teaching activities, thus becoming another instrument for mediating between the subject, their world and scientific knowledge. To do this, we need to experiment and theorize a lot about science education, with one eye on the past and the other on the future, but above all with our conscience alive in the present (GIORDAN, 1999, p. 49).

Taking the use of technologies in educational processes as a point of reference, the following is a reflection on Information and Communication Technologies (ICTs) and the teaching of chemistry.

1.3.1.2 Technologies in chemistry teaching

"Technologies are as old as the human species" (KENSKI, 2007, p. 15). This sentence counters the idea that technology is an advent of modernity. The author explains that it was precisely human creativity, in its different eras, that allowed the development of all the technologies that are available today. She also states that technology is power, harking back to the Stone Age, when man began to master artifacts in order to impose control over nature and other men.

Now, what is technology? The answer to this question is relative. Depending on the different philosophical conceptions that each individual has, each view of it can be different. Some may say that technology is a set of science-based advances that man has mastered for his own benefit. Others may say that technology is an accumulation of discoveries that humanity has made, but that it does not exercise dominion, but is dominated by the technologies imposed by contemporary times. Perhaps others may have different ideas from these and still understand what technology is and what it represents today.

Cupani (2004) describes a philosophically-based reflection on technology. He presents the conceptions of three scholars: Mario Bunge (analytical approach); Albert Borgmann (phenomenological approach); and Andrew Feenberg (critical philosophy approach). In common, each of these perspectives presents an understanding of what technology is, considering its relationship with man and society.

Bunge adopts a view of technology that is directly related to scientific knowledge, but points out that it cannot be reduced to the use of this type of knowledge. Cupani (2004) reports that

Bunge understands technique as the control or transformation of nature by man, who makes use of pre-scientific knowledge. Technology, in turn, consists of scientifically-based technique (CUPANI, 2004, p. 3).

Andrew Feenberg presents a theoretical basis on the philosophy of technology, highlighting social and philosophical aspects such as power, who dominates and who is dominated by technology, efficiency and technological rationality. He states that

Technology is a two-sided phenomenon: the operator on the one hand, and the object on the other. Where operator and object are human beings, technical action is an exercise of power. Where, moreover, society is organized around technology, technological power is the main form of power in society (FEENBERG, 2005, p. 49).

For Borgmann, technology is a way in which man deals with the world, through a pattern interwoven with everyday life. He emphasizes the importance of devices and how they are inserted into our lives.

Technology becomes more concrete and evident in (technological) devices, in objects such as television sets, central heating plants, cars and the like. Devices therefore represent clear and accessible instances of the standard or paradigm of modern technology (BORGMAM,

1984, p. 3).

But to what extent should people see technology as the solution to problems, especially those related to education? Compared to a medicine, it's not enough for it to exist, it has to be administered correctly and sometimes several of them are indicated for the same treatment. It's also true that it doesn't always restore health to the person who used it. Many technologies exist, but how well are they being used in education? To what extent can technology help the teaching process?

Sampaio and Elia (2012) state that

Proposals for innovation, because they have seductive appeals, since they often involve the use of ICTs accompanied by promising speeches about their beneficial effects in solving educational problems, tend to hide the complex reality of this process. This is because, whether vertical or horizontal in nature, every change causes a certain degree of uncertainty and insecurity, since it interferes with institutionalized habits and routines, arousing different attitudes and perceptions (SAMPAIO & ELIA, 2012, p. 36).

This work aims to provide contributions so that these uncertainties and insecurities can be mitigated within the context of teaching practice. It is possible to put ICTs in their place within educational processes. Not as the essence of the teaching and learning process, not as an end in themselves, but as tools that can help achieve the goal of education. It is possible to teach without ICTs, but it cannot be denied that they bring new possibilities that can help educational processes and teaching practice. Benite and Benite (2008) point out that we live in a time of great advances in communication between peoples through communication and information technologies. Although this is a global and widespread phenomenon, it is education that is being most affected by this technological wave. However, they warn that

When you depend on something that you don't understand how it works, you run the risk of becoming its slave. Hence the importance of computerizing the classroom being preceded by a reflection that involves all the participants: teachers, students, teaching and administrative staff and the community in a careful training process. This avoids dependence on technology (and future disillusionment), wasted time, human and financial resources. It is essential that computerization is "thought out" based on the specific political-pedagogical project for a given school (BENITE & BENITE, 2008, p. 14).

Just like the definition of technology, ICTs can be understood in different ways, and may or may not be associated with educational environments. However, the definition expressed by Tavares and collaborators is interesting and clear when they say that ICTs are "a set of technological resources that can provide communication and/or automation of various types of processes in various areas and especially in teaching and research" (TAVARES et al., 2013, p. 156). They also say that schools with access to technologies such as computers can use them as a tool in the teaching and learning process.

Giordan (2005) warns that if computers are used to strengthen education, they will be subject to a complex system of cultural, political, economic and fundamentally educational relationships. Kenski (2007, p. 53) also says that "the new digital technologies do not offer their users a new world, without problems". In other words, technologies, especially educational ones, do not represent salvation and liberation from the bottlenecks faced by citizens and educators. On the other hand,

The inclusion of ICT in education can be an important tool for improving the teaching-learning process. These technologies can generate positive or negative results, depending on how they are used. However, any new technique can only be used with ease and naturalness after a long process of appropriation. In the case of ICTs, this process clearly involves two facets that it would be a mistake to confuse: the technological and the pedagogical (LEITE & RIBEIRO, 2012, p. 4).

Educational processes can be transformed for the better and technologies can play an important role in this. However, they cannot be the end goal of teaching and learning processes, especially in primary education. Education is the end, technologies are some of the possible means to achieve it.

As Benite and Benite (2008) state, it is necessary for schools to integrate ICTs into their practices, as it is clear that these technologies are already widespread in our society. These authors also suggest that schools can act as compensators for the social and regional inequalities that unequal access to ICTs is generating.

For Melo and Melo (2005), it is a mistake to take the computer as a tool that will solve problems related to education on its own. According to them, this has frustrated teachers, especially chemistry teachers, with regard to the use of new technologies and their importance in the teaching-learning process. This misconception must be overcome (or alleviated) if ICTs are to be permanently included in chemistry teachers' didactics. To do this, it is necessary to look at this context in terms of learning and development processes.

Santiago (2010) and Gabini (2005 and 2008) indicate that it is the teachers who are the subject of research into teaching chemistry with a focus on ICTs. In fact, it is the teachers who are the main protagonists in the teaching and learning process, as they are the points of reference for the students, they are the ones who propose one methodological practice or another, and they are the ones who are held responsible for the students' failures.

It is also understood that there would be no teacher without students, but investing in training and stimulating those who carry out the function of teaching means investing in and stimulating the training of the students themselves. It is agreed that digital resources, especially the *internet and software*, are basic elements for teacher training, both initial and ongoing (SERRA, 2009).

In this sense, Tavares and collaborators (2013) bring up a series of important reflections on the role of the teacher within a school context where ICTs are inserted. For them,

[...] In order to promote the use of computers in schools as a teaching-learning process, it is necessary for teachers to have a knowledge of computers that they can use as a source of learning [...].

[...] However, teachers do not need to be computer experts, but they do need to have a reasonable knowledge of the area, as they must know how to use *software* that facilitates the transmission of knowledge and the work of students. [...]

[...] For teachers who have difficulty using information and communication technologies, it is appropriate for them to seek new knowledge through courses and continuing education. [...] (TAVARES et al., 2013, p. 158 and 159).

Giordan (2005) reflects on the transformations that computers have brought to teachers' work. Unlike other activities, where computers have replaced many human jobs, this has not happened in teaching. For him, there is a greater demand for this professional today than in times past due to population growth, and not because of some technological "miracle".

However, Santiago (2010) states that state school teachers in Manaus (AM) did not receive adequate specific instruction on the use of ICTs in chemistry teaching during their undergraduate studies or even in the continuing education courses offered by public bodies. The same author observed that the most commonly used technological tools were televisions and computers to prepare lessons, but not in the classroom.

The results of Gabini (2005) also confirm that, if well planned and executed, pedagogical activities permeated by digital resources have a great chance of being successful. Breaking out of the traditional lecture routine can already make students more interested in the lessons, making them more attentive and willing to learn. However, even if the lessons are not essentially lectures, it is important that alternative activities to this type of methodology are planned and developed with a focus on the student and their learning.

The results that Gabini (2008) presents in his thesis regarding the poor use of spaces for continuing education are worrying. Unfortunately, it is known that laboratories, computers, *tablets* and other materials made available by the Department of Education or by the school itself are often not used by teachers for various reasons. This situation was also reported by Santos and Azevedo (2012).

Although many schools have these technologies, they are not used as they should be, often being locked away in isolated rooms and far from the use of students and teachers, because they are unable to connect these resources to teaching activities (SANTOS & AZEVEDO, 2012, p. 3).

Gabini (2008) notes that teachers want to change their practices, but many of these desires don't materialize in the classroom for various reasons. However, a greater incentive for teachers to overcome their anxieties, accompanied by an efficient continuing education program and improvements in digital spaces in schools, are fundamental ingredients for ICTs to be more present in chemistry classes.

Tavares and colleagues (2013) extol the potential of technologies in teaching chemistry. However, for them it is necessary for ICTs to be used appropriately, so that they can provide the student with a broader view of the subject studied, enabling a better understanding of the content, without neglecting the student's reality. In this way, knowledge mediated by technology can facilitate the didactic transposition of the content studied in Chemistry, transforming superficial learning into meaningful learning.

1.3.2 Learning theories and ICTs

1.3.2.1 How does the individual learn?

Understanding how an individual learns and develops is essential for any educational activity, especially those carried out in a school environment. Understanding a student's characteristics within social, biological and psychological contexts obliges every teacher to reflect on their own teaching practices. Learning or not learning certain content involves a series of factors, not just cognitive ones. It requires a broad reflection on the school, family, social and individual context of each student. In this sense,

We know that education is a scenario of diversity, both in terms of students and teachers, made up of a complexity of problems. As such, we can't entirely focus on the student's lack of interest or on their family and the environment in which they live, because there are professionals who are blocked from educational innovation, which is necessary to try to minimize learning problems (SILVA, 2012a, p. 3).

The path to efficient learning depends, as has already been observed, on many factors. The teacher's action, the student's intellectual type, the opportunities offered by the school's immediate environment, and even the student's life interests. Because of this, the school can no longer be seen as a literacy machine. It is hoped that students will be led to a critical education capable of understanding and transforming the environment in which they live. Considering the school context, where each individual is expected to learn and develop their potential. Lopes (2000) states that

Day-to-day school life has shown us some very questionable situations in this regard. First of all, the educational objectives proposed in course curricula are confusing and disconnected from social reality. The content to be taught, in turn, is defined in an authoritarian way, as teachers, as a rule, do not take part in this task. Under these conditions, they tend to lack meaningful links with the students' life experiences, interests and needs (LOPES, 2000, p. 41).

The current situation of basic education, especially secondary education, in Brazil is very worrying. Recently, the magazine Isto É (ISTO É, 2013, p. 52-56) published a survey showing that half of all students between the ages of 15 and 17 are not enrolled in secondary school. Among the various reasons given to justify this reality, we can highlight the year/grade gap and the inadequacy of the school to the life, expectations and needs of young people.

Nardi (2009) states that schools "kill" children's natural curiosity to learn and discover something new. This "murder", according to the author, is caused by poor, outdated teaching that falls short of the real needs and interests of young people. So far, one question seems to have emerged: how can we help to make students more interested in the knowledge that is taught in schools? The answer to this very important question is certainly not dissociated from the very understanding of how human beings learn and how they learn best. The immeasurable educational contributions of Piaget and Vygotsky are guiding points for designing a more comprehensive, interactive and interesting education for any subject in the teaching and learning process.

Today's technological development   has taken root in all age groups in society. However, mastering certain computer skills, such as access to social networks and basic handling of operating systems, may not be enough to guarantee the use of ICTs as educational tools. Even with this technological level that society and schools are currently living at, it should be noted that

For the teaching/learning process to be successful, there needs to be interaction between the student and the teacher, so that the teacher and his or her pedagogical practice provide the conditions for the student to explore his or her knowledge and to be able to reconstruct it by interacting with each other in the search for solutions to problems. In addition, it is important to restore the students' enjoyment of studying by valuing creativity and thought, by introducing the use of new methodologies in the search for motivation for more meaningful learning, seeking to promote a constructivist environment with integration between teachers and their teaching content, in an attempt to avoid boredom, complacency, disinterest and discouragement (CABRERA & SALVI, 2005, p. 9).

In this respect, the works of Jean William Fritz Piaget (1896 - 1980) and Lev Semenovich Vygotsky (1896 - 1934) are the foundations for discussing theories of learning and development. Despite the notable differences between the conceptions of these two thinkers, one cannot fail to emphasize that both conceived of the child (or young person) as an active being who interacts with the environment that surrounds them.

These questions point to the need to reflect on the foundations of Piaget's and Vygotsky's theories and understand how they permeate the context of ICTs in chemistry teaching, especially in public schools.

1.3.2.2 Jean Piaget and genetic epistemology

The main concepts of Piaget's Learning Theory, presented below, are based on the book "Intelligence and Affectivity of the Child in Jean Piaget's Theory" by Wadsworth (1995). Jean Piaget developed a theory of learning called Genetic Epistemology, in which he sought to understand the development of human intelligence and the process of knowledge, using the clinical method.

Piaget's main focus, according to Wadsworth (1995), was on the child's behavior and how it interacts with the environment, objects and other people. In this respect, he adopted the interactionist line, as it overcame mechanistic materialism and its educational implications. Thus, learning requires interaction.

However, perhaps influenced by his training in biology, Piaget (WADSWORTH, 1995) states that intellectual activity cannot be seen in isolation from the biological functioning of the organism of the thinking being. Schema, assimilation, accommodation and equilibration are the four fundamental concepts to explain the cognitive development of human beings.

•   Schema: These are mental or cognitive structures by which individuals intellectually adapt to and organize the environment;

•   Assimilation: This is the cognitive process by which a person integrates new perceptual, motor or conceptual data into existing schemes;

•   Accommodation: Accommodation is the creation of new schemas or the modification of old schemas, resulting in a change in the cognitive structure or its development;

•   Equilibration: This is a self-regulating process whose instruments are assimilation and accommodation, allowing external experience to be incorporated into the internal structure (schemas).

Adopting the thesis that development takes place continuously, not abruptly, Piaget (WADSWORTH, 1995) dared to categorize the phases of cognitive development into four stages:

•   Stage of sensorimotor intelligence (0 - 2 years);

•   Pre-operational stage of thought (2 - 7 years);

•   Concrete operations stage (7 - 11 years);

•   Formal operations stage (11 - 15 or more).

In this respect, reflecting Piaget's ideas, Chakur (1995) states that:

Knowledge essentially consists of operative processes that transform reality, either in actions or in thoughts, in order to understand the mechanism of these transformations and thus assimilate events and objects into systems of operations (or structures of transformations) (CHAKUR, 1995, p. 45).

Taking a pedagogical look at these ideas defended by Piaget, knowledge is not just transmitted, but is constructed by the subject's own activity, and the school needs to be seen as a place of exploration and discovery. In this way, the teacher is a facilitator and stimulator of the whole teaching and learning process.

Piaget and the Teaching of Chemistry

Learning and development are fundamental to the structuring of educational processes. According to Wadsworth (1995), Piaget argues that learning takes place in an environment around children (interaction) and that it only becomes evident after the individual's organic development. Considering the age ranges of the stages of development presented by Piaget (1998), it can be seen that students attending secondary school (generally between 15 and 17 years old) are already at the stage of formal operations (1115 years old or more).

Thus, by developing interactive activities based on ICTs, it is hoped that students will have the ability to reason logically in order to overcome the challenges proposed in chemistry classes. Maia and colleagues (2013) say that chemistry requires students to be at a formal level in order to understand the concepts.

Piaget (1998) states that to educate is to adapt the individual to the social environment. This statement suggests that in order to educate today, in today's technological society, it is important to consider the use of available communication and information technologies in teaching and learning situations.

In this sense, the importance of the teacher cannot be reduced, since they can choose how to teach their students. For Souza and Bezerra (2013), Piaget himself points out that the aim of teaching is to promote the development of citizens capable of acting on new things and not just repeating what past generations have done. Being creative, innovative and a discoverer are virtues expected of any individual. In this context, ICTs can be excellent tools in the teaching and learning process.

Piaget (2010) reflects critically on schools and their tendency to promote student development based on the transfer of content and adult authority. He advocates an active pedagogy in which the student takes on the role of rediscoverer or reconstructor of knowledge. In his words:

How many schools talk about development, interest, spontaneous activity, etc. when, in reality, it's just about the development provided for in the syllabus, forced interests and activities suggested by adult authority. The real criterion for an active pedagogy (a form of education that is as rare today as it was in the 17th century) is based, it seems to us, on the way in which truth is acquired: There is no authentic activity as long as the pupil accepts the truth of a statement only because it is transmitted by the adult with all the force of the explicit or implicit authority of the teacher's word or the text of the manual; there is activity, on the contrary - when the pupil rediscovers or reconstructs the truth on the basis of material or internalized actions - which consist of experimenting or reasoning for himself (PIAGET, 2010, p. 24). 24).

Silva (2012a) discusses a work entitled "Piaget for Chemists" by Herron (1975). Again, it is clear that the content taught in chemistry requires students to be at the formal operational stage. Still focusing on Piaget and the Teaching of Chemistry, the author presents the implications of Piaget's theory for the teaching and learning process, stating that the pedagogical objectives should be geared towards the interests of the students and that the content worked on should only serve as a process of natural evolution, not having an end in itself.

At the stage of formal operations, an important concept in Piaget's theory, presented by Wadsworth (1995), is reflective abstraction. Defined as one of the mechanisms by which the process of cognitive construction takes place, this abstraction is essential in chemistry teaching due to the need to understand abstract logical-mathematical knowledge. One of the great advantages of ICTs is their ability to visually present concepts that were previously only imagined mentally, making it easier to reach more abstract cognitive levels. For example, the three-dimensionality of spatial isomerism is often difficult for students to work with. However, this can be overcome by using technologies such as physical or digital structural models.

In this regard, Souza and Bezerra (2013) state that the development of digital spatial reasoning contributes to the ability to observe, assimilate and understand spaces and objects. However, it's not enough to insert technologies into this development; it's essential that the student is interested and can see and understand the notions of three-dimensionality, whether in digital spaces or not. In fact, if students are not interested and want to learn, it will be difficult for them to learn what they are trying to teach, even with the use of ICTs.

Carvalho (1983), in a paper entitled "Piaget and the Teaching of Science", warns against trying to simplify Jean Piaget's complex theory. The author also states that formal reasoning is directly related to science teaching. In addition, she argues that:

[...] We can't limit ourselves to teaching only at the level of specific knowledge, because we wouldn't be giving our students a chance to develop either intellectually or even within our own content. [...]

[...] We cannot fail people just because they are not yet capable of formal reasoning. [...]

[...] It is possible that, as well as teaching chemistry, we are giving them the opportunity to develop intellectually. [...] (CARVALHO, 1983, p. 76).

Wadsworth (1995) answers the following question: "Is the use of computers compatible with Piaget's theory?". Answering this question is not easy, since Piaget's theory predates the popularization of computers. However, some reflections can be made.

Computers, machines, don't build knowledge for children, for example, you don't learn to calculate on a calculator. Operating a calculator is different from building knowledge of mathematical operations. However, even though they are not sources of knowledge in themselves, computers are machines that have the potential to involve children, stimulating them to act. In this action, in this interaction, a lot of knowledge can be built up more quickly and dynamically.

1.3.2.3 Lev Vygotsky and the cultural-historical approach to cognitive development

The conceptual approaches of Lev Vygotsky's ideas are based on his book "The Social Formation of the Mind" (VYGOSTSKY, 1998). The genesis of his theory lies in the intrinsic relationship between the psychological profiles of the human being and the socio-cultural environment in which the individual is inserted. For him, development is mediated by physical or symbolic instruments, and the social aspect is responsible for the higher psychological processes.

Vygotsky (1998) defines these processes as functions that control conscious behavior, intentional action and the individual's freedom in relation to the characteristics of the present moment and space. These higher psychological functions, unlike the elementary ones (reflex, simple and elementary), coordinate various internal activities, such as: the ability to think about absent objects, to solve problems and to conceive new concepts.

In this context, Vygotsky (1998) describes four guiding concepts for understanding psychological issues:

•   Phylogenetic: Development of the human species;

• Ontogenetic: Development of the individual;

•   Sociogenetic: History of social groups;

•   Microgenetic: Development of specific aspects of the subjects' psychological repertoire.

The psychological development of each individual, for Vygotsky (1998), is peculiar to each human being, being a continuous and variable process and not something universal. In this sense, an important concept is that of symbolic mediation, i.e. a process of intervention of an element in a relationship. Thus, any indirect activities are mediated by the action of signs (internal tools) and instruments (external tools).

The concept of internalization is understood in Vygotsky's theory as a process of reconstruction, which begins with external operations, but which are then internalized. This process is not trivial, but is followed by a series of transformations. These changes can be described as follows: an operation that initially represents an external activity is reconstructed and begins to occur internally; an interpersonal process is transformed into an intrapersonal process; and the transformation of an interpersonal process into an intrapersonal process is the result of a long series of events that have taken place over the course of development (VYGOTSKY, 1998).

Perhaps the best-known concept in Vygotsky's theory is the Zone of Proximal Development (ZPD). It characterizes the distance between the level of actual development and the level of potential development, determined through the solution of problems under the guidance of an adult or in collaboration with more experienced companions. What characterizes the level of actual development is mental development in a retrospective way, while the ZPD designates mental development in a prospective way. This zone allows us to understand the functions of development that are under construction.

Pedagogically, the fundamental concepts of Vygotsky's work have important implications. Firstly, the child must be seen as an interactive being and that interactions with other individuals culminate in the construction of internal meanings. In this context, it is necessary to consider who learns, who teaches and the relationship between them, and it is up to the teacher to participate effectively in the process by mediating learning. The teacher must encourage the student to reflect on what is implicit in what is explicit. Here, the subject of the teaching and learning process is the student-teacher-medium group.

Still on the subject of Vygotsky, Cachapuz and colleagues (2004) state that

The meaning of a given concept (or situation) is the result of interaction with others (teacher or students), mediated through language, which is the means by which students are encouraged to reflect and explain, in order to understand how their experiences and contextualized knowledge are integrated into a broader system (CACHAPUZ et al., 2004, p. 16).

The same authors also reflect on Vygotsky's ideas and the role of the teacher. For them, it is important to value the social life in the class, the varied interactions, as well as the mediating role of the teacher in the construction of students' knowledge (CACHAPUZ et al., 2004).

Cunha (2009) reflects on perception in Vygotsky's approach. For her, the emphasis is on the processes of using the higher functions of thought, mediated by the symbolic and sociocultural representation of these processes. She also states that,

When we perceive elements of the real world, we relate these perceptions to our information, which is present in the psychological apparatus. The perceived object is perceived as a complete entity and not as a pile of information captured by the senses. This is related to the individual's development path, their knowledge of the world and their experiences (CUNHA, 2009, p. 35).

Considering that symbolism is essential to Chemical Knowledge, as it deals with many representations (formulas, equations, structures, etc.), the structuring concepts presented in Vygotsky's work are consistent with the complexity inherent in understanding Chemistry. In the same way, the stages presented by Piaget lead us to understand why, in schools, chemistry is generally studied in the final years of primary education.

Vygotsky and the Teaching of Chemistry

Souza and Bezerra (2013) state that the growth of ICTs, combined with the diversification of strategies for their use, has the potential to promote reflection on teaching practices and can stimulate active learning. With regard to Vygotsky, Kohl (2000) describes that, unlike the Piagetian conception, learning is the engine of development. Thus, the more you learn, the more you develop. Based on this principle, we understand that the school should be able to leverage learning, but reality often testifies against this need.

Maia and colleagues (2013) argue that, according to Vygotsky's conceptions of learning, the teacher's work should be centered on the Zone of Proximal Development of their learners, facilitating the students' contact with the environment that promotes interaction and, consequently, learning. It can be seen that the teacher's role is crucial to the success of the students' learning process, since on their own, they would not have the same interactions and, therefore, their learning would be compromised. In this context, the

The teacher becomes responsible for creating zones of proximal development, i.e. providing conditions and situations for the student to transform and develop a more significant cognitive process in their mind (CABRERA & SALVI, 2005, p. 3).

According to Kohl (2000), Vygotsky positions the relationship between man and the world, mediated by symbolic systems, as the theoretical basis for his ideas. In this relationship, we cannot conceive of the term "world" in isolation from the technological systems present in people's daily lives. Based on this analysis, it should be noted that ICTs can be present in the symbolic and representational media situations characteristic of chemistry teaching, facilitating learning.

Vygotsky (1998) describes the process of internalization as the mental incorporation of external facts. The evolutionary process of symbolic (and also instrumental) use is similar to the development of language itself. ICT resources are a set of signs, instruments and language that aim to facilitate the process of internalization, in the case of this master's research, internalizing the chemical knowledge taught in secondary school.

Santiago (2010) warns that failure to bring the digital world into the classroom can make the teacher's job more difficult, and can also hinder the efficiency of the teaching and learning process. It should be emphasized, however, that technologies should not be blamed for the failure of the teaching and learning process, nor should they even be blamed for the failure of the teaching and learning process.

all the credit in the event of success. In this sense, more important than the technologies is the person who can manipulate them: the teacher.

In a very simplistic analysis, Piaget focuses his interactionism on the subject/physical object relationship, but Vygotsky's conception is centered on the interaction between the individual and others with more experience. Although different, the two views do not contradict each other. For Gabini (2008), it is necessary to consider the actions of the subjects who manipulate computer technology resources. The teacher's role in this respect is to guide (mediate) the use of ICTs, but the student themselves must also "venture out" and interact with the technologies themselves.

As for the role of the teacher, Santos (2004) considers that access to, use of and mastery of new information and communication technologies are inherent needs for all those who have the task of teaching. This idea is in line with what Vygotsky expects of the teacher, an agent who helps the students' metacognition process through prospective teaching.

1.3.3 Teaching Organic Chemistry and the potential of ChemSketch software

1.3.3.1 Organic Chemistry in High School

The National Curriculum Parameters for Secondary Education (PCNEM, BRASIL 1999) were drawn up with the aim of helping school teams to carry out their activities, guiding the production of state documents, such as the Curriculum Guidelines for Secondary Education in the state of Acre. As described in the National Curriculum Guidelines for Secondary Education, current curricular practice is as follows

remains predominantly disciplinary, with a linear and fragmented view of knowledge in the structure of the disciplines themselves, despite numerous experiments carried out as part of pedagogical projects influenced by the Parameters (BRASIL, 2006, p. 101).

The proposals put forward by the PCNEM defend a vision of education based on the role of the student as an active citizen who is aware of their actions. However, actions based on the transfer of content and the passivity of the student in the teaching and learning process are still dominant in the current reality of basic education in Brazil. The PCNEM states that,

The extreme complexity of today's world no longer allows secondary education to be just a preparation for a selection exam, where the student is an expert because they are trained to solve questions that always require the same standard answer. Today's world demands that students take a stand, make judgments and decisions, and be held accountable for doing so (BRASIL, 2006, p. 103).

This need for students to position themselves, make judgments and decisions shows that individuals need to be aware of the social, environmental, political, economic and religious factors that structure the society in which they live. In this respect, it is understood that chemical knowledge cannot be treated in a superficial way; on the contrary, this implies that the students need to be aware of the social, environmental, political and decision-making factors that structure the society in which they live.

[...] understand the chemical transformations that take place in the physical world in a comprehensive and integrated way, so that they can make informed judgments about information from cultural tradition, the media and the school itself, and make decisions autonomously as individuals and citizens (BRASIL, 2006, p. 106).

Thus, according to the Educational Guidelines Complementary to the National Curriculum Parameters (PCN+), it is believed that high school chemistry can

[...] be an instrument of human formation that broadens cultural horizons and autonomy in the exercise of citizenship, if chemical knowledge is promoted as one of the means of interpreting the world and intervening in reality, if it is presented as a science, with its own concepts, methods and languages, and as a historical construction, related to technological development and the many aspects of life in society. (BRASIL, 2002, p. 87).

The Curriculum Guidelines for the state of Acre are in line with the PCN+, as they also place the understanding of content as a means of understanding the transformations evident in practical life, as well as providing the means for students to have the ability to solve chemical problems faced in their daily lives (ACRE, 2010). However, to what extent has the school today formed the reflective citizens envisioned by the national documents and state guidelines? Does the school offer the human formation that really broadens students' cultural horizons and citizen autonomy? Reflecting on these questions, it is clear that basic education in Brazil still falls far short of what its own guidelines envisioned and that many challenges remain to be overcome.

Now, what are the challenges for the teacher in the classroom? The Curriculum Guidelines list a series of points on this issue. These are

Take as a principle that, in a classroom, the most important thing is the students.

To consider that leadership, dialogue and reflection-action are fundamental in the management of pedagogical work.

Build and consolidate, as far as possible, explicit and shared projects with the students.

Making the local dimension - the specific needs of the class - and the general dimension - the demands of the school's educational project and the education system - compatible in pedagogical work.

Ensuring the exercise of citizenship in everyday classroom life.

Articulate, in teaching action, the perspective of teaching and class management.

Create contexts that encourage students to play a leading role.

Encouraging students to develop an appropriate student attitude and a commitment to their own learning.

Producing knowledge about what happens in everyday life, seeking answers to challenges - whenever possible, collectively.

Consider the classroom and the students to be 'open systems', i.e. they are in permanent interaction with everything beyond themselves and the classroom door (ACRE, 2010, p. 12-13).

The ten points highlighted as the teacher's challenges in the classroom seek to value the figure of the student and their daily life. Although the challenge for teachers to adapt their practices to the context of today's technological society has not been directly exposed, it is assumed that the use of ICTs should also be a challenge for teachers. The challenge of considering and, if appropriate, using available technologies that can improve the learning and interest of students, as well as the teacher's own work.

With regard to the teaching of Organic Chemistry, a more critical look at the guidelines in the Acre Curriculum Guidelines shows that there is still a vivid understanding of its content. Regarding Organic Chemistry, it is stated that

It is essential to know how to construct structural formulas, name compounds and identify organic functions. Among the various existing functions, the most important, mainly because of their presence in everyday modern life, are hydrocarbons, alcohols, aldehydes, ketones, carboxylic acids, esters, ethers, phenols, nitro compounds, amines, amides and organic halides. The study of foods, especially macronutrients - carbohydrates, proteins and lipids - can provide a better understanding of the transformations that take place in the body, as well as making it possible to calculate the energy value of nutrients. The study of isomerism can be complementary and is important for differentiating between similar compounds. The study of some organic reactions, mainly those of addition, substitution and esterification, with a focus on reaction mechanisms, completes the theoretical exploration of organic chemistry. Industrial applications of these reactions can be discussed, particularly the production of plastics such as polyethylenes and PETs, as well as their consequences for the environment, related to their chemical resistance (ACRE, 2010, p. 24).

The initial focus on the skills of constructing formulas, naming compounds and identifying organic functions should not be worked on in a content-based way. However, it is known that often when working on these subjects, teachers basically use the textbook and solving exercises as teaching methodologies. Practices like this may not provide students with the conditions to be reflective and see the real importance of this specific content in their lives.

Something worrying about the proposals for activities suggested in the Acre Curriculum Guidelines (ACRE, 2010) is the fact that there is no proposal to use ICTs for chemistry teachers, with the exception of suggested research that can use the internet. In other words, if a teacher follows "to the letter" the suggestions presented in the Acre guidelines for teaching Organic Chemistry, he or she can work on all the content for the 3rd year without using any ICT. Of course, a teacher can promote the learning of Organic Chemistry without the use of *software* and other technologies, but the guidelines should at least suggest to teachers which technologies could possibly be used in the classroom. In this respect, it can be seen that the Curriculum Guidelines do not adequately address the work of teachers, nor do they mention the importance of technology in teaching chemistry in Acre.

The curriculum guidelines for Acre present a set of teaching objectives, emphasizing what is expected of students on completion of the 3rd year of secondary school:

Develop knowledge of organic chemistry through the study of hydrocarbons, identifying their presence in the contemporary world.

Recognize, name and represent the formulas of compounds of different organic functions (oxygenated, nitrogenated and halogenated) and understand the role and effects of the different compounds present in food, neurotransmitters, licit substances (medicines and alcoholic beverages) and illicit substances (psychoactive drugs) in the human organism.

Distinguish the different planar and spatial isomers in terms of structural formulas, names and properties, equate the chemical transformations of organic compounds for the manufacture of different consumer products and understand the molecular structures and manufacturing processes of different polymers present in industrialized products in our daily lives (ACRE, 2010, p. 26).

For these objectives to be achieved, it is important that students are interested in studying and want to learn what they are being taught. The use of ICTs, such as the *ChemSketch software*, in the teaching of content related to the objectives mentioned above, can facilitate the teaching and learning process, as well as increasing students' interest in the subject. However, *ChemSketch*, or any other *software,* will not "work miracles" and drastically transform the teaching of Organic Chemistry, much less replace the role of the teacher in this process, but it can help in the understanding of the contents of this important area of Chemistry. The possibility, for example, of transforming molecules into virtual, three-dimensional images can make it easier for students to learn, since the difficulty of the representational level of chemistry can be mitigated by bringing abstract objects closer to concrete ones.

1.3.3.2 Molecular structuring software

Not all *software* available for computers and portable devices can be classified as educational *software*. For it to be, a given program must be developed taking into account the teaching and learning context and its pedagogical aspects. However, even if it is designed as educational, it is necessary for the teacher to adapt this tool to their teaching methodology. In this way, the most important thing is that there is a homogenization between *software* and practice, in such a way that programs are integrated into the educational context, not the other way around (LEITE, 2015).

It should also be noted that "the effectiveness of educational *software* depends on the role assigned to it and the pedagogical articulation assigned by the teacher" (LEITE, 2015, p. 177). Leite (2015) points out that programs that purport to be educational need to have a theoretical and pedagogical foundation, focusing on interaction and ease of use for students.

There are many educational programs aimed at teaching chemistry, some of which allow the creation and manipulation of molecular structures. Some molecular structuring *software* and its main features will be presented below. *ACD/ChemSketch* will be detailed later.

ChemDraw

ChemDraw Professional (Figure 3) is robust molecular structure *software*. It was developed in the 1980s by David A. Evans and Stewart Rubenstein and can be downloaded from https://scistore.cambridgesoft.com/Chemistry/Chemical_Structure_Drawing. *ChemDraw* is part of PerkinElmer's *ChemOffice* package, along with other resources.

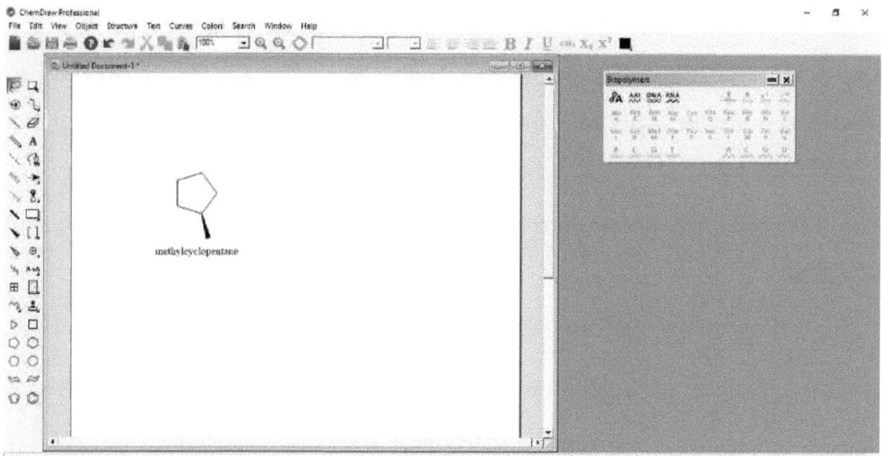

Figure 3 - *ChemDraw Professional* interface.

Source: *print screen of* the program on the Windows 10 operating system.

Aimed mainly at researchers in Organic Chemistry, this program has the following features, among others:

- Determine the nomenclature from the chemical structure;

- Determine the chemical structure from the name of the compound;

- Simulate nuclear magnetic resonance spectra (NMR[1] H and[13] C);

- Simulate mass spectra;

- Calculate the stoichiometry of chemical reactions;

- Access various structural design models;

- Convert drawn structures into three-dimensional models;

- Access organic structure databases from a web browser.

However, because *ChemDraw* is designed for high-quality structural chemical drawing, it is a robust program that requires computers with good *hardware* configurations to run without crashing, which can make it difficult to use in the classroom context in basic chemistry education. It has a two-week free *trial* and a paid *pro* version, the latter of which is more complete.

BkChem

Another *software program* for drawing organic structures is *BkChem* (Figure 4). This free *software* can be downloaded from http://bkchem.zirael.org/. Its main functions are:

•	Draw compounds taking into account the bonding capacity of atoms and their angles;

•	It has ready-made models, such as organic rings;

•	It allows the use of radicals, arrows and other basic graphic resources;

•	Allows structures to be rotated in 2D and 3D;

Its simple interface is not very intuitive, and drawing relatively simple structures can be more difficult than in other programs that perform the same function.

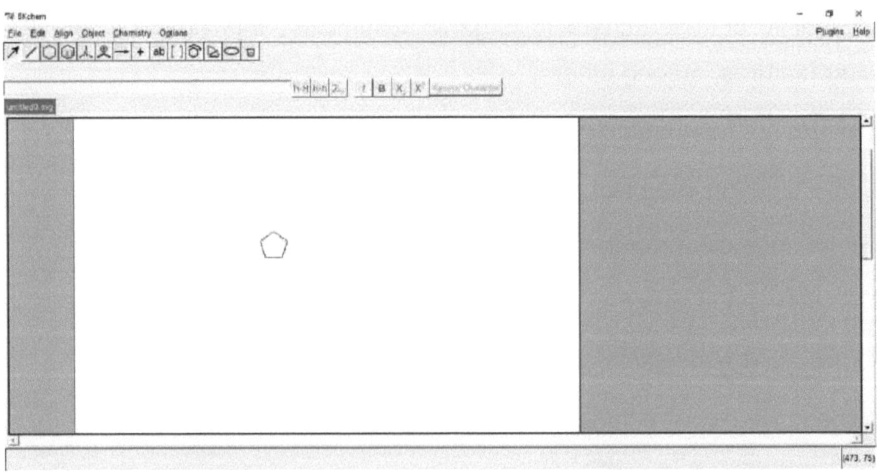

Figure 4 - *BkChem* interface.

Source: *print screen of* the program on the Windows 10 operating system.

ChemWindow

KnowItAll ChemWindow Edition (Figure 5) is *software* developed for drawing chemical structures, more geared towards scientific research. It can be downloaded from http://www.bio-rad.com/en- us/product/chemical-structure-drawing-software?pcp_loc=catprod.

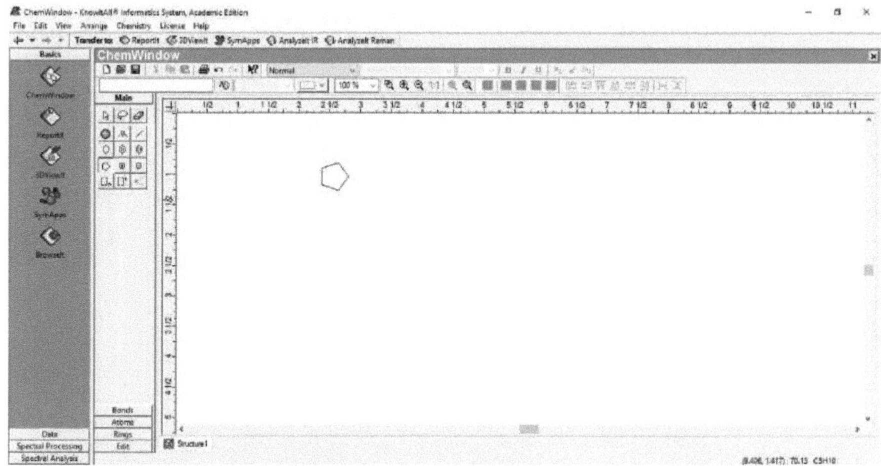

Figure 5 - *KnowItAll ChemWindow Edition* interface.

Source: *print screen of* the program on the Windows 10 operating system.

The main disadvantage of this program is that it doesn't have a free version, as is the case with *ChemSketch* and *BkChem*. Its main features are:

• Draw 2D structures and visualize them in 3D.

• It has a database with structures, information on chemical properties, etc;

• Calculates bond lengths, angles, etc;

• It has tools for drawing rings, atoms, electrons, charges, chains, arrows and more;

• Stereochemical recognition including R / S and E / Z isomers;

• Capable of generating names for organic structures.

Symyx Draw

Symyx Draw 4.0 (Figure 6) is a molecular structuring program (a replacement for *ISIS/Draw*) and is free for personal, non-commercial use. It is currently developed and marketed by the company BIOVIA, through its website http://accelrys.com/. The 4.0 version of *Symyx Draw* can still   be downloaded from the website:

https://copy.com/yyfjPZzJg2PHylcw/SymyxDraw-4.0_AE.zip?download=1.

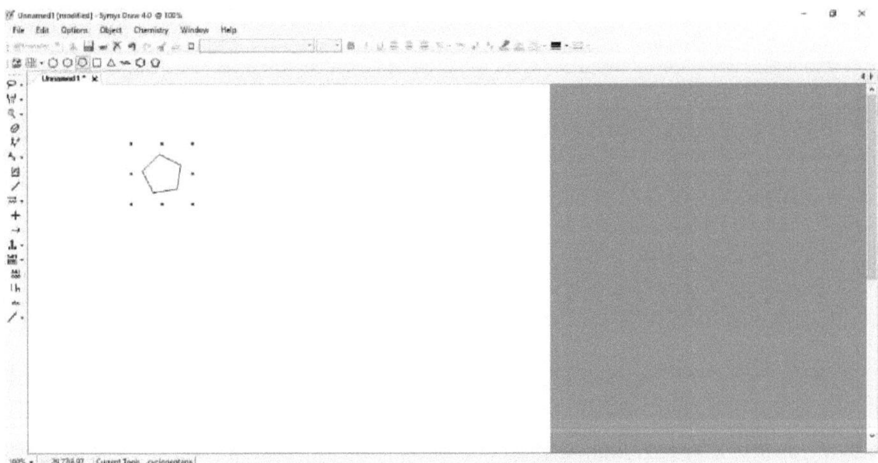

Figure 6 - *Symyx Draw 4.0* interface.

Source: *print screen of* the program on the Windows 10 operating system.

Its main functions are:

- It allows the use of dynamic models to create complex reactions;

- Creates structures with R groups for queries or enumerations;

- It is compatible with *Microsoft Office*;

- Converts names to structures and structures to names;

- Allows the visualization of molecules in 3D;

It should be noted that, like *ChemWindow* and *ChemDraw*, *Symyx Draw's* features are more geared towards scientists, chemists and other researchers who need to create organic structures and manipulate them. Its interfaces hinder more educational applications, such as the creation of structures by high school students.

1.3.3.3 ChemSketch

ACD/ChemSketch freeware (Figure 7) is molecular structuring *software* from *Advanced Chemistry Development Inc*, which can be downloaded from http://www.acdlabs.com/resources/freeware/chemsketch/. It has many features that can be used in chemistry teaching situations, from secondary to higher education.

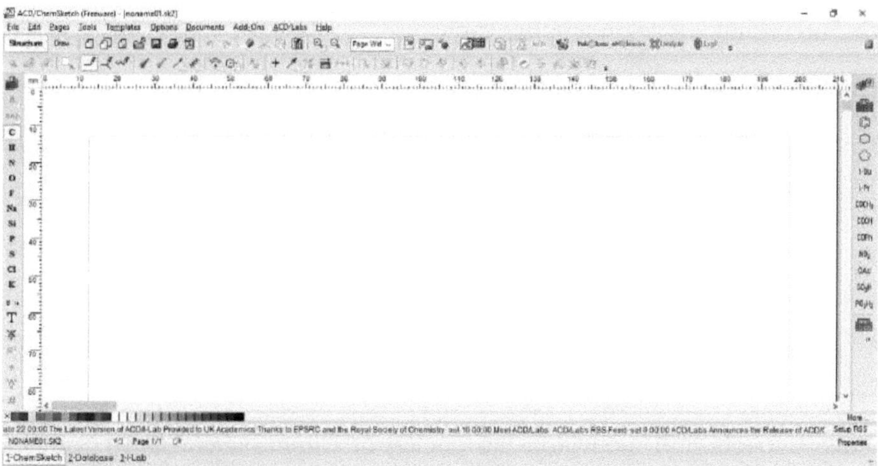

Figure 7. *ChemSketch freeware* 2015 interface.

Source: *print screen of* the program on the Windows 10 operating system.

It allows you to draw chemical structures, including organic structures, organometallic compounds and polymers. Its tools include the possibility of:

•   Assemble flat structures and optimize them for three-dimensional visualization;

•   Manipulate 3D structures;

•   Name, according to the rules of the International Union of Pure and Applied Chemistry (IUPAC), structures of up to fifty atoms and three cycles;

- Access a database of *templates* with various ready-made structures of different classes of compounds (alkaloids, vitamins, carbohydrates, nucleic acids, aromatics, among others);

- Determine the stereochemical data of the structures drawn;

- Determine various properties of the compounds drawn, such as: molecular formula, molar mass, density, surface tension, tautomeric forms, refractive index, molar volume, etc...;

- Save your projects in various formats: jpeg, png, gif, pdf, among others, in addition to the program's standard format;

- Set up organic reaction mechanisms;

- Numbering carbons in a carbon chain;

- Consult online databases for articles, chromatographic and spectroscopic data, medicinal applications and other information on the structures designed.

As you can see, *ChemSketch* has functionalities that are directly related to the Organic Chemistry content taught in the 3rd year of high school in Acre, and can help teachers when preparing their assessments, presentations and other activities.

The main aspects that motivated the choice of *ChemSketch*, and not other molecular structuring *software*, as the object of research for this master's degree were:

- Availability of a free version with most of the program's features;

- Simple and intuitive interface;

- Compatibility with *Microsoft Office* programs;

- Variety of extensions and options for salvaging structures;

- It is a lightweight program that works even on computers with low processing power;

- Possibility of determining the properties of the structures created;

- Easy visualization and manipulation of 3D structures.

Demonstrations of how to use the program's main tools are detailed in the *Practical Guide to Using ChemSketch*, the product of this research. *ChemSketch* has two versions, one free (*Freeware*) and one commercial. Leite (2015) shows the main differences between these versions in Table 1.

Table 1. Differences between the commercial and free versions of ACD/Labs ChemSketch

Function	Commercial	Free of charge
Advanced drawing tools	X	X
Identifying tautomers	X	X
Names of structures with up to 50 atoms	X	X
3D viewer	X	X
Export to Adobe PDF	X	X
ACD Dictionary	X	
Search for files on your computer by structure	X	
ACD/Labs extensions for *ChemDraw*	X	
Technical Support	X	

Source: LEITE (2015).

A study by Li and colleagues (2004) compared the functionalities of four molecular structuring *software programs*: *ChemDraw*, *ChemWindow*, *ISIS/Draw* (predecessor of *Symyx Draw*), and *ChemSketch*, in their most current versions at the time. Although this work is more than ten years old, the main functions of the programs evaluated are still the same, although some improvements have been made to each of them. Table 2 shows an evaluation of these four *software programs* by these authors.

Table 2. Personal experiences with the four types of software Chemistry

Software	ChemDraw	ChemWindow	ISIS/Draw	ChemSketch

2D drawing	Excellent	Good	Excellent	Excellent
3D	Excellent	Good	Medium	Good
Laboratory	Medium	Excellent	ND	Good
Spectres	Medium	Medium	ND	Excellent
Online functions	Excellent	ND	Excellent	Excellent
User programming	ND	ND	ND	Yes
Supported file formats	Excellent	Medium	Medium	Medium
Scope	Good	Medium	Medium	Excellent
Overall	Excellent	Good	Good	Excellent

Source: LI et al. (2004).

ND = not available.

Considering the authors' evaluation, it can be seen that *ChemSketch* was the *software* that received the best evaluation, followed by *ChemDraw*. The same authors also point out that all these programs have their virtues, and the choice of which is best depends on the user's purpose. However, they suggest that *ChemSketch is* the most notable due to its overall performance and intelligent development.

One of the possibilities of *ChemSketch* is to work on the representational level of chemistry. Visuospatial skills (derived from the concept of spatial visualization) are important for students to be able to move between the different aspects of the representational level of chemistry. Being able to manipulate models and use model-building tools is an important factor in the development of visuospatial skills.

In Chemistry, this ability is essential, as learning involves visuospatial skills that support performing certain cognitive operations spatially. Through these operations, students can become capable of internalizing external visualizations and then manipulating the structures mentally.

For example, if a class was asked what methane (CH_4) is, some might answer that it is a gaseous substance, usually derived from oil or the decomposition of organic matter. It would also be correct to say that methane is one of the main components of natural gas and that it is also present in biogas. Others could talk about its importance for the world's energy matrix and for the production of plastics. However, how could these students internally abstract the definition of methane to a representational chemical level? What does a single methane molecule look like? One of the possibilities for teachers to work on this question is to use visualization *software* such as *ChemSketch*.

This program allows the visualization and animation of representational models in both 2D and 3D, as shown in Figure 8. These representations can also be viewed as automatic or manual animations, depending on who is manipulating the program.

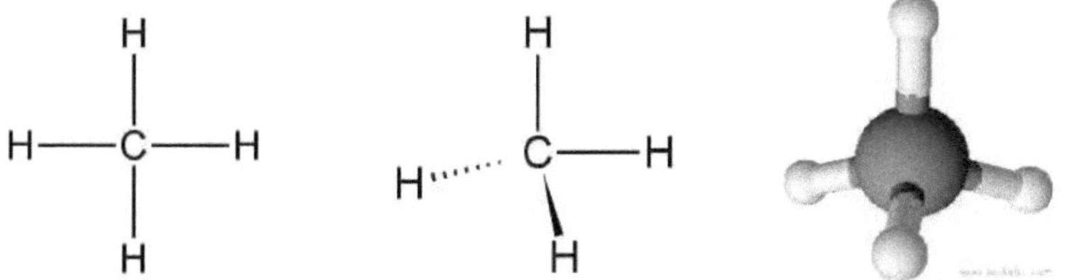

Figure 8. Different representations of methane. Source: Prepared by the author.

The use of computer tools, such as *ChemSketch*, can facilitate understanding of the representational level of chemistry, as well as motivating learning and helping students adapt to an increasingly technological society. It is clear that this *software* can be an integral tool in the teaching practices of chemistry teachers. It can also be used by the students themselves to solve exercises, draw and manipulate organic structures, among other possibilities.

2 OBJECTIVES

2.1 General

The main objective of this project is to contribute to the process of teaching Organic Chemistry in Basic Education, through the use of the ACD/Labs *ChemSketch Freeware software*, considering its potential and limitations, in the context of the state of Acre.

2.2 Specifics

> To study the *ChemSketch software* in detail, in order to learn about its features and how they can be used in high school organic chemistry teaching situations;

> To produce a Practical Guide to using *ChemSketch* to teach Organic Chemistry, as well as other content, for chemistry teachers in Rio Branco - Acre;

> To promote the *software* and the Practical Guide in a mini-course for chemistry teachers in Rio Branco;

> Contribute to the ongoing training of primary school teachers in Rio Branco.

3 METHODOLOGY

The research entitled THE USE OF THE CHEMSKETCH SOFTWARE AS A TOOL IN THE TEACHING OF ORGANIC CHEMISTRY IN BASIC EDUCATION IN THE STATE OF ACRE was carried out in four stages described below:

1ª Stage: In-depth study of *ChemSketch* software

ChemSketch freeware, in its 2015 version, is a *software program* with many features that can be used when studying certain chemistry subjects, especially those related to organic chemistry. In this initial phase, an in-depth study of the program's tools was carried out, based on the support documents offered by *ACD/Labs* itself, as well as video tutorials and other sources.

This stage was fundamental to mastering the features that *ChemSketch* offers, allowing us to evaluate which features are most important, considering the Organic Chemistry content studied in High School.

Stage 2: Preparation of a Practical Guide to Using *ChemSketch* to Teach Organic Chemistry (product)

The **Practical Guide to using *ChemSketch*** (preliminary version) was produced with the aim of making it easier for chemistry teachers and other interested parties to use the *ChemSketch software*. The preliminary version of the Practical Guide consists of nine chapters, spread over ninety-two pages. In the Guide, the main functions of the program have been described in an illustrated way, with a focus on objectivity in the exposition of these functionalities.

3ª Stage: Dissemination and evaluation of the Practical Guide to using *ChemSketch* through a mini-course

After preparing the Practical Guide to using *ChemSketch*, a mini-course was held to disseminate and evaluate this product. The mini-course, entitled ***ChemSketch*: Learning to Draw Organic Structures**, was held on October

21 and 22, 2015, as part of the Viver Ciência program, promoted by the State Department of Education and Sport (SEE-AC). The event was aimed at chemistry teachers working in primary education in Rio Branco - Acre.

During the mini-course, participants received and evaluated the Practical Guide using questionnaires, as well as contributing to the research by suggesting improvements to the product and presenting proposals for didactic activities using *ChemSketch* to teach chemistry in primary education. In addition to collecting data through questionnaires, audio recordings were made with the consent of the mini-course participants, in order to record suggestions and criticisms debated during the event.

Stage 4: Preparation of the final version of the Practical Guide to Using ChemSketch

Taking into account the suggestions and criticisms described in the questionnaires and in the suggestions for activity proposals submitted by the participants of the mini-course **Learning to Draw Organic Structures**, the **Practical Guide to Using *ChemSketch*** has been improved and expanded with the inclusion of a chapter dedicated to explaining the suggested proposals.

The development of the aforementioned stages was fundamental to clarifying the proposed research question: "How can the *ACD/ChemSketch Freeware software* be used as a media tool in the process of teaching Organic Chemistry in Secondary Education, considering its potential and limitations, in the context of Basic Education in the state of Acre? ". The preparation of the Practical Guide, based on the knowledge

in-depth study of the *software*, the guidelines for teaching organic chemistry in Acre and the contributions of the teachers who took part in the workshop, we tried to see how the *ChemSketch software* could be useful in the process of teaching chemistry in basic education in Acre.

4 RESULTS AND DISCUSSION

4.1 Short course report

Initially, the proposal to disseminate and evaluate the Practical Guide to Using *ChemSketch* was to hold a continuing education course with chemistry teachers from the Rio Branco public school system, with the support of the State Department of Education and Sport (SEE-AC). However, due to the strike movement that took place in 2015, all the continuing education courses in Chemistry scheduled for the second semester were postponed until 2016. The training offered was planned to last a minimum of twelve hours.

However, this situation led to a change in the planning of this master's project. Instead of offering an official SEE-AC continuing education course, a mini-course was held as part of the Viver Ciência program, also in partnership with this secretariat. The event was open to the whole community, but the target audience was primary school chemistry teachers.

Teachers were invited to attend via websites, printed material, e-mails, social networks and personal conversations. Due to the time constraints of the event, the mini-course had to be held in just eight hours.

The following is general information and a descriptive account of the mini-course.

Minicourse General Information

Short course: CHEMSKETCH: LEARNING TO DRAW ORGANIC STRUCTURES

Speaker: Alcides Loureiro Santos

Date/Time: October 21 and 22, 2015 from 2:00 p.m. to 5:00 p.m.

Location: NIEAD Coordination Room - NTI Block - UFAC

Number of participants: 13

Report

1st DAY OF MINICURSE - October 21, 2015

The mini-course began with an oral presentation by the lecturer, who told the participants the name of the mini-course, that it was part of the Viver Ciência (SEE-AC) program and that it was also part of a research project for the Professional Master's Degree in Science and Mathematics Teaching (MPECIM-UFAC). Each participant then introduced themselves, saying their name and the institution they were affiliated with (Figure 9).

Figure 9. The mini-course. Source: The author.

It turned out that the workshop's audience was made up of seven chemistry teachers who work in primary education, three chemistry undergraduates, a recently graduated master's student in chemistry, a physics teacher and a forestry engineer. The latter participated only as a listener, not answering the questionnaires or signing the informed consent form.

After introducing the participants, a space was opened for each one to say something about the topic: **The use of technology in chemistry teaching**. The participants were told that their speeches would be audio-recorded. The first participant, a chemistry teacher, to explain the proposed topic emphasized the importance of simulating chemistry and that technology contributes greatly to teaching, since not everything can be demonstrated in the laboratory. This

46

thought agrees with Melo and Melo (2005), when they state that

The apprehension of new knowledge through new technologies must be contemplated and, since the school is the place par excellence for promoting new learning, it is necessary to include tools such as simulation software in the teaching methodology. An additional way of using computers in teaching is to use them to simulate experiences. Instead of the student observing reality in a fragmented way, as in chemistry and physics laboratories, experiments are simulated on the computer screen (MELO & MELO, 2005, p. 4).

Another chemistry teacher talked about the public investment in equipment, which was not always used effectively by teachers and students. She also talked about the possibility of using internet resources, such as virtual periodic tables to investigate the properties of chemical elements. In fact, the use of ICTs in chemistry teaching is increasingly being researched and drawing the attention of teachers because of its educational potential (MELO & MELO, 2005, p. 8).

A chemistry undergraduate pointed out that, in the case of the *netbooks* distributed to third-year high school students in 2011 in the state public school system, they helped students prepare for the entrance exams and that there were good tools that could be used. He also emphasized that the problem was not the computers that were distributed, but rather their misuse, and that the teachers themselves, due to a lack of adequate training, did not guide the students on how to use these technologies productively.

This concern about teacher training, raised by a chemistry undergraduate, shows that future teachers are already reflecting on the quality of their initial training at university level. This concern also appears in the statements of practising teachers about their continuing education. With regard to this scenario, Borges (2010) points out that

teacher training is limited to imitating "successful" projects, which disregards the reflective capacities and strategies of each teacher. The definition of 'what', 'how' and 'why' we teach is indicated by researchers, disregarding the intentionality of the process. There is no dialogue between those who define and those who apply the concepts. Reflection often focuses only on the teacher-student relationship, without considering the external influences that shape the course of the lesson and the teacher's training (BORGES, 2011, p. 18).

After this, the lecturer gave a presentation on the Prezi platform which covered: the objectives of the mini-course; what *ChemSketch Freeware* 2015 is; how to obtain the program's installation file; an overview of the *software* and its main features. Afterwards, a preliminary version of the **Practical Guide to Using ChemSketch** was distributed free of charge to each participant.

The lecturer read out the TFIC, explaining the objectives of the research entitled "THE USE OF THE CHEMSKETCH SOFTWARE AS A TOOL IN THE TEACHING OF ORGANIC CHEMISTRY IN THE BASIC EDUCATION OF THE STATE OF ACRE" and the free participation of each person in the research. After distributing the ICF, the participants were given the Initial Questionnaire. After signing the form and answering the questions, a break was taken.

Upon returning to the mini-course, *ChemSketch* was presented in the sequence of the chapters in the Practical Guide. The first six chapters were presented:

1. WHAT IS CHEMSKETCH?

2. PROGRAM INSTALLATION

3. WAYS OF WORKING

4. BARS, TEMPLATES AND TABLES

5. CREATING AND EDITING STRUCTURES

6. EXPLORING STRUCTURES

At the end of the first day of the mini-course, the lecturer explained the plans for the second day and what activities would be carried out.

2nd DAY OF THE MINICOURSE - October 22, 2015

On the second day of the mini-course, the presentation of the Practical Guide was resumed, together with an explanation of the main functions of

ChemSketch. The remaining chapters were worked through:

7. 3D VISUALIZATION AND MANIPULATION

8. CHEMSKETCH DATABASE

9. DRAW

The mini-course participants were asked to assemble some organic structures, such as lapachol (Figure 10). As the functions described in the Practical Guide were presented, some doubts were raised by the participants, which were clarified together. After this stage, it was explained that the last chapter of the Practical Guide would be built on proposals for activities suggested by the participants themselves.

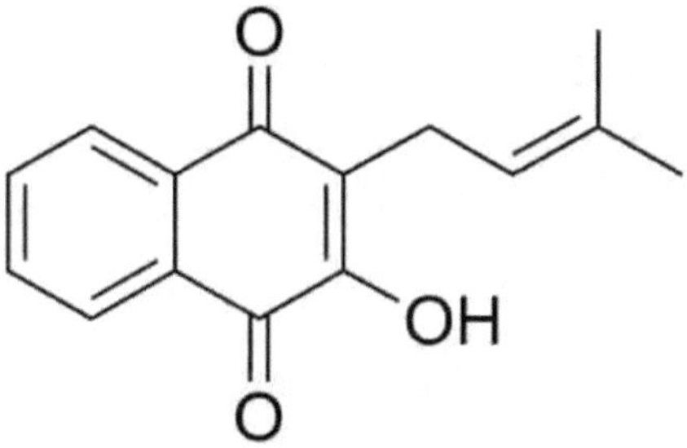

Figure 10. Structure of lapachol.

Source: Prepared by the author.

After the break, the participants, using the knowledge and skills presented in the mini-course and in the Practical Guide itself, shared some suggestions for didactic activities using the *ChemSketch software.* Seven suggested activities were presented, which will be described in more detail later.

One participant reported that more teachers had been invited and would have liked to take part in the mini-course. However, due to the school coordinators not giving their permission, these teachers were unable to attend the event. It was said that, because it was being held during the week of the ENEM (National High School Exam), the school coordinators claimed that the students were in the final stretch of preparation for the exam and that, for this reason, the teachers could not be released.

Other participants also reported the difficulties and lack of support that public school teachers have in taking part in training and courses, even though they are required to innovate in their teaching practices. In this sense, in order for the integration of ICTs into chemistry teaching, as a differentiated and innovative methodology, to be implemented in schools, "teacher training and pedagogical preparation are necessary" (SANTIAGO, 2010, p. 23).

The lecturer asked if any of the participants had any more suggestions for activities using *ChemSketch*. One participant said that, after getting to know the tools, this *software* could be used in all chemistry subjects, especially organic chemistry and in demonstrations of experimental activities.

It is also worth noting that one participant suggested producing a summary manual, a pocket guide, for *ChemSketch*. This would present only the essential functions of the program and would be easier to consult than the Practical Guide distributed.

Finally, the participants answered the Final Questionnaire, evaluating the mini-course and the **Practical Guide to Using *ChemSketch***, as well as suggesting an answer to the research question for this master's degree. The lecturer then thanked everyone for their participation and the mini-course ended.

4.2 Initial Questionnaire Results

Twelve people took part in the mini-course and the survey, using the questionnaires and activity proposals, as described above in the event report. Some questions will be analyzed considering the responses of all the participants, while others will be discussed according to the performance of the mini-course participants.

To facilitate the presentation of the results and their discussion, the audience will be divided into three groups:

1. **Group 1 (G1):** Current chemistry teachers - 7 participants

2. **Group 2 (G2):** Chemistry graduates and teachers who are not in the classroom - 4 participants

3. **Group 3 (G3):** Acting physics teacher - 1 participant In this sense, the first question of the questionnaire was the following:

4. How important do you think the use of Information and Communication Technologies (ICT) is for teaching chemistry today?

( ) Very important

( ) Important

( ) Not very important

( ) Not important

Considering the responses of groups G1 and G2, as can be seen in Graph 1, the importance of ICT in teaching chemistry is unanimous. While agreeing with this relationship of relevance, it is pertinent to highlight the challenging role that teachers have in teaching contextualized classes that offer new teaching technologies (OLIVEIRA, 2010).

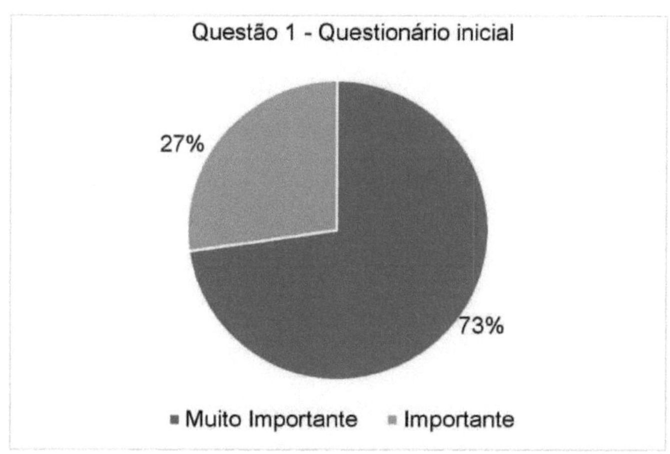

Graph 1. Results of Question 1 - Initial questionnaire (G1 and G2).
Source: Prepared by the author.

Santiago (2010) reflects on the theme of ICTs in chemistry teaching, stating

51

that:

ICTs provide the chemistry teacher and the student with access to a wide variety of scientific information: regional, national and international, making it possible to create methodologies and strategies for the teaching-learning process in chemistry, so that the teacher and the student can construct chemical knowledge in a contextualized way, giving the student a true understanding of why they study this subject in high school (SANTIAGO, 2010, p. 47).

Question 2 of the Initial Questionnaire read as follows:

2. Do you use chemistry software in your classes?

( ) Yes   ( ) No

If so, which ones do you use the most?

As this is a question about the use of *software* in chemistry classes, it is coherent that only Group G1 is considered in this discussion. The result is shown in Graph 2.

It's important to see that, even though they are already working as chemistry teachers in primary education, some with more than ten years' experience, the majority of teachers still don't use chemistry *software* in their classes.

Looking again at the answers to Question 1, it can be seen that, even considering the importance of using ICTs, teachers may still not use them in their practice. According to Santiago (2010), this is because many teachers have had poor initial and continuing training.

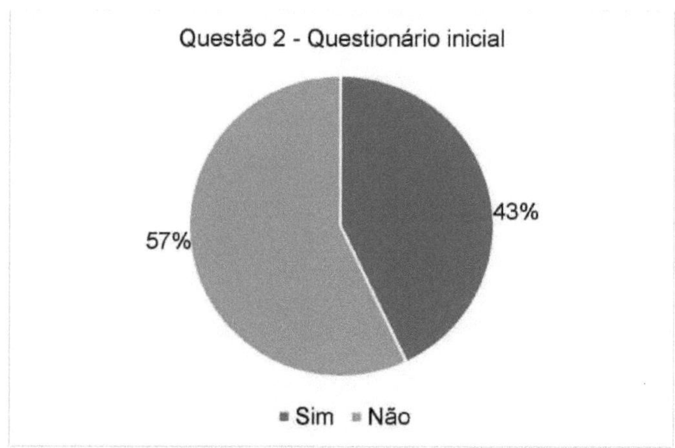

Graph 2. Results of Question 2 - Initial questionnaire (G1). Source: Prepared by the author.

Considering all the participants, only four of them mentioned *software* used in their classes. They mentioned *ChemSketch*, *Phet Simulations*, virtual laboratories and interactive periodic tables. It was noted that only one chemistry teacher had ever used *ChemSketch* and that no other molecular structuring *software* had been used by the participants in their classes.

The third question in the initial questionnaire asked:

3. In your classes, which chemistry subjects would you like to use software developed to stimulate your students' learning?

Graph 3 shows which and how many times chemistry content was mentioned by groups G1 and G2.

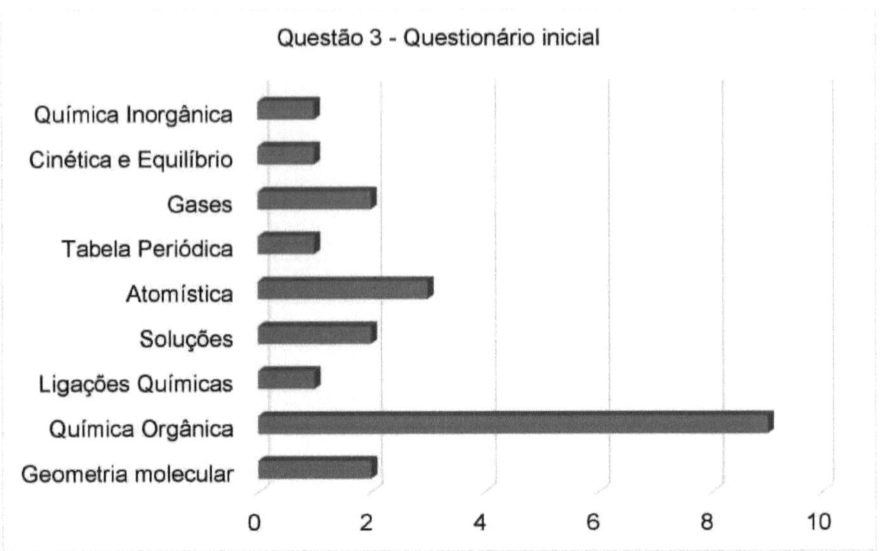

Graph 3. Results of Question 3 - Initial questionnaire (G1 and G2). Source: Prepared by the author.

The results obtained indicate that Organic Chemistry is the content that teachers would most like to use in their classes. In this way, *ChemSketch* corresponds to this desire of teachers, since it is essentially designed for applications of organic content.

Raupp and colleagues (2010) discuss the role of computer representations in chemistry teaching, stating that *software* can help students learn. This is because students' understanding and performance in certain chemistry subjects is related to their capacity for three-dimensional abstraction (PAVLINIC et al., 2007).

The fourth question of the first survey asked:

7. Have you taken part in any continuing training courses aimed at introducing ICTs for teaching chemistry?

Only group G1 was considered, since they already work in primary education. Four of the seven teachers said they had never taken part in continuing training on ICT. This is a worrying situation, given that teachers are expected to take on different and innovative positions in the classroom, but they are not always prepared to offer teaching that goes beyond the traditional.

Even though the difference between the ages of teachers and students is not even one or two decades, the speed with which technologies are being created and integrated into society's way of life further reaffirms the need for continuing training based on ICTs. In this sense, Santiago (2010) reports that:

We are living through a time of transition in which teachers in general were born into an analog technological world, unlike students who were born into the digital age. Not being able to bring this digital world into their classroom makes their work more difficult and the teaching and learning process does not take place in a meaningful way (SANTIAGO, 2010, p. 39).

In this context, Silva (2012b, p. 39) states that "it will be essential for teachers to demonstrate their potential in order to present not only the importance of ICTs, but also how they can be appropriated in teaching/learning". In addition to initial and continuing training, several aspects need to be considered in order for teachers to meet the expectations of current scientific, technological and social education. The same author highlights some of these aspects:

In order to reach this level where teachers have technological knowledge, it is necessary to overcome challenges, including: training or updating professionals, increasing the level of digital literacy in the country, minimally qualifying new technical and higher education professionals, significantly increasing the training of specialists in new technologies at all levels in all areas of new technologies, making large-scale use of new information and communication technologies in distance learning. Universities should therefore update their training matrices with subjects focused on education and ICTs (SILVA, 2012b, p. 41).

The next question asked:

5. Do you know the ChemSketch software?

( ) Yes   ( ) No

Considering groups G1, G2 and G3, it was observed that half of the mini-course participants already knew *ChemSketch*. One participant said she knew the *software* "more or less". If we relate this result to the third question in the initial questionnaire, we can see that the fact of knowing a piece of *software,* in the case of Organic Chemistry, is not enough for teachers to use it in the classroom with their students. The lack of continuing training in the use of ICTs for teaching chemistry may explain this situation.

It's not enough to know the tool; teachers need to be trained and have mastery and confidence in it, so that they can decide whether and how to use it in their teaching practice (SILVA, 2012b).

In the sixth question, the participants answered:

6. How would you rate your level of mastery of ChemSketch?

( ) Very

( ) Fair

( ) Not much

( ) None

Graph 4 shows the results of this question:

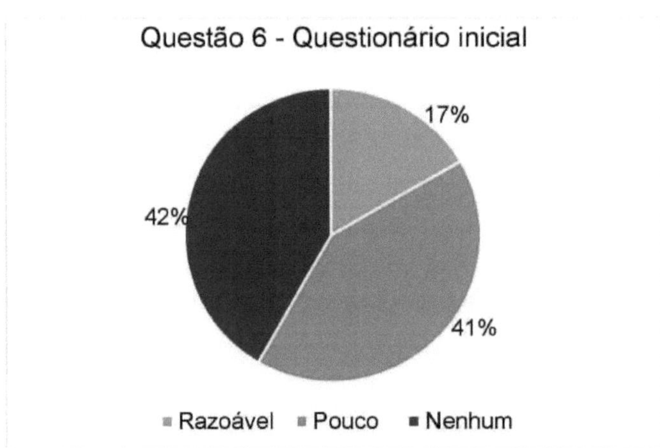

Graph 4. Results of Question 6 - Initial questionnaire (G1, G2 and G3). Source: Prepared by the author.

The low level of knowledge of *ChemSketch* on the part of the participants makes it clear that there is a need for more initiatives to disseminate and provide teachers with knowledge of chemistry educational *software.* It should be noted that there are several other molecular structuring programs, but the accessibility and ease of use of *ChemSketch* suggests that it could be integrated into initial and continuing chemistry training. This would probably fulfill teachers' desire to use educational *software* and organic chemistry.

The seventh and final question of the initial questionnaire was the following:

10. Can molecular structure programs contribute to your teaching practice in the classroom and help in the process of teaching and learning chemistry?

( ) Yes   ( ) No   ( ) Depends

Would you like to comment?

All the participants answered "yes", agreeing that molecular structuring *software* can help their practice and contribute to students' learning. Most of the participants commented on this question, some of which are presented below:

Comment 1: There is no doubt that the program contributes efficiently to teaching practice, as it facilitates the learning of organic chemistry, among other subjects, in various ways.

Comment 2: Tools like *ChemSketch* contribute a lot to the teaching and learning process. So it's up to us teachers to look for different and innovative methodologies.

These two comments emphasize that *ChemSketch* can contribute to the process of teaching and learning chemistry. It is important to note that in comment 2, the role of the teacher was highlighted in the search for differentiated and innovative methodologies.

Other comments were:

Comment 3: All knowledge is valid, even more so using digital resources that enable interaction between the student and teacher in the classroom.

Comment 4: Given the complexity of the chemistry content, any initiative to encourage student participation is welcome.

Comment 5: Making the lesson interesting for the students, facilitating learning.

Comment 6: It makes the content more attractive to the student, so that it is taught widely.

Generally speaking, in comments 3 to 6, one figure was highlighted: the student. The interactivity of the classes, the greater interest of the students and a greater attraction to the content were the main points highlighted. In this way, it is understood that molecular structuring *software*, such as *ChemSketch*, are tools that have the potential to contribute to the teachers' way of teaching and to the goal: that the student learns.

Learning requires interest. If you're not interested in a particular subject, you won't see the need to learn it. The adoption of new technologies and more interactive and participatory classes make it more effective for students to learn, as they are generally more interesting than traditional classes, based on teachers teaching the content. In other words, it is necessary to overcome an ideology characterized by teaching by transmission/reception , towards a ideology in which the student learns through discovery, interaction and acting on their own learning (GIORDAN, 2008).

Group member G3, the physics teacher, pointed out that:

Comment 7: Yes, in some cases it is practically impossible to demonstrate the structure. In these cases, the program is essential.

In fact, demonstrating organic structures, especially the more complex ones, is a difficult task using only traditional resources: textbooks and blackboards, for example. In this respect, the virtualization of structures, objects and experiments can be an excellent tool in the process of teaching and learning science (RAUPP et al., 2010).

4.3 Final Questionnaire Results

The final questionnaire was answered at the end of the mini-course on October 22, 2015. The first question was precisely the Research Question of this master's degree, namely:

1. How can the ACD/ChemSketch Freeware software be used as a media tool in the process of teaching Organic Chemistry in High School, considering its potential and limitations, in the context of Basic Education in the state of Acre?

All the participants in the mini-course answered the question. The answers given by the three groups of participants are highlighted below.

G1 Group

Answer 1 : The *software* can be taken into school practice, it can be made available to chemistry and biology teachers, who can work on a lot of content, facilitating learning in geometry, organic chemistry, laboratory materials, isomerism, among others.

Answer 2: The *ACD/ChemSketch software* can be used for various organic chemistry subjects, such as determining the nomenclature, geometry, isomerism and reaction mechanisms of organic compounds. In this sense, it is possible to apply activities to students that involve demonstrating the structural formula of known compounds using the program, as well as detailing the reactions (step by step) that lead up to obtaining the compound.

Answer 3: Preparation of tests for secondary and higher education. Students making molecules using the program. Games using data from the *ChemSketch* program. Separation methods: in which the teacher describes the methods and the student creates the drawing of the structure needed to separate the mixture; creating a graph for thermochemistry, among others.

Answer 4: The *software* can be used in various areas of teaching (chemistry, physics and biology). In organic chemistry, it can be used to help construct compounds, mechanisms and isomerism.

Answer 5: The program can be introduced in public and private schools, and in education departments as an alternative tool for teaching chemistry.

Answer 6: The *software* is great, it can be used in many ways, not to mention that if it is applied as a lesson in the laboratory, it can be a great tool for learning.

Answer 7: In drawing up teaching action plans. Like workshops to bring all the subjects together. Work on different subjects such as chemistry, physics, biology and English.

The responses from the teachers in group G1 suggest that *ChemSketch* can be used positively in teaching and learning situations in chemistry, especially in organic chemistry content. The positive evaluations of the *software* reinforce its educational characteristics.

Some teachers reported that, in addition to Organic Chemistry, *ChemSketch* can be used for other subjects, such as the separation of mixtures, thermochemistry and basic laboratory concepts.

As presented in the proposals suggested by the teachers, *ChemSketch*'s resources can help with practical activities in the laboratory. In this sense, Teruya and colleagues (2013) report in their review article that some authors highlight:

the cost of maintaining a laboratory and the large number of students in classes as factors that would encourage the use of computer tools, allowing students to simulate different techniques and procedures, as well as using different reagents, which would not always be possible in a real laboratory environment. The use of simulations is also reported for distance learning situations, with the aims of familiarizing students on a distance learning chemistry course with the laboratory and preparing them for face-to-face practical classes. As pre-laboratory activities, simulations can also help to better prepare students for practical work by training them in techniques and providing the theoretical background associated with the experiment. This would make laboratory activities much more meaningful for students (TERUYA et al., 2013, p. 565).

These statements show that the program is not only useful in the third year of high school, when organic chemistry is usually studied, but also in the first year (separation of mixtures, for example) and the second year (thermochemistry). In fact, as Raupp and colleagues (2010, p. 32) state, "using this type of *software* can help students not only with isomerism problems, but with a wider range of chemistry problems".

It is important to note that some teachers reported other subjects, such as biology, physics and English. In this respect, the possibility of working in an interdisciplinary way could be another way of attracting students' interest in lessons, making it easier for them to learn.

As for group G2, the answers were:

Group G2

Answer 8: Always using it in organic classes makes it much easier in the classroom, learning is very important.

Answer 9: Using a playful way to draw the student's attention to the subject.

Answer 10: I believe that all teachers have computers, but if they don't, most schools make them available for use, either at school or at home. Therefore, the possibilities are many, because the program is simple/easy to access, it is up to each professional to find out about it and apply it in their daily practices, because in addition to facilitating the teaching process, it will possibly reflect on learning.

Answer 11: To make it easier for the students to understand.

In general, it is possible to see that the answers given by the participants in group G2 were more generic than those in group G1, since no specific activities were mentioned. Even so, it is important to stress the importance given to the *software* in the study of Organic Chemistry. In addition, it was said that the program can facilitate student learning. Playful activities involving *ChemSketch* are also possibilities for attracting students' attention to chemistry lessons.

Group G3

Answer 12: The application could be used in practical classes in a science laboratory, where students could assemble the structure on the computer and then set it up in a laboratory as a physical model that could be visualized in a practical way.

The spontaneous participation of a physics teacher in the mini-course was very interesting. Even though the *software* is geared towards chemistry, he was able to see the potential of *ChemSketch* as a media tool for teaching and learning science. In fact, the possibility of three-dimensional molecular modeling, the visualization of biomolecules, the creation of graphs, the assembly of laboratory schemes, show that *ChemSketch* is an inter and multidisciplinary educational *software.*

Regarding the Practical Guide to Using *ChemSketch*, the second question in the final questionnaire asked:

2. *What rating would you give ChemSketch's user guide?*

(   ) *Good*

(   ) *Good*

(   ) *Regular*

(   ) *Bad*

(   ) *Bad*

Graph 5 shows the results of this question for the three groups of mini-course participants.

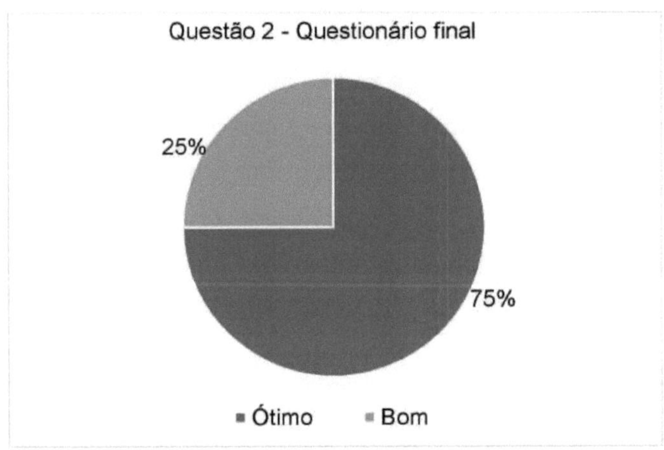

Graph 5. Results of Question 2 - Final questionnaire (G1, G2 and G3). Source: Prepared by the author.

The excellent evaluation given to the Practical Guide shows that, even though the *software* is in English, providing practical and illustrative material can make it easier to master digital tools such as *ChemSketch*. Each participant in the mini-course received the material free of charge and did not have to return it at the end of the event. This result confirms the positive validation of *ChemSketch* as a product of this master's research.

However, in order to obtain a more specific evaluation, two more questions about the *ChemSketch* Practical Guide were included in the final questionnaire.

3. What did you like most about the Practical Guide to using ChemSketch?

4. What did you like least about the Practical Guide to using ChemSketch?

Table 3 shows the answers to questions 3 and 4 of the final questionnaire, considering all the groups of participants in the mini-course. Not all participants answered these questions, and the third question received more responses than the fourth.

Table 3. Answers to questions 3 and 4 of the final questionnaire

3. What you liked most...	4. What you liked least...
Practicality in naming structures, assembling structures and researching properties.	The limitation of the number of atoms for naming structures.
The fact that it details each of the program's toolbars.	The size of the letters in the pictures
Detailed information.	Image size
Explanatory form.	English
Clarity, facilitates program functions	There should be more content
Great, well explained and illustrated.	
The *prints*, practicality.	
Easy to explain.	
Easy and self-explanatory language.	

Source: Prepared by the author.

One answer from the third question and two from the fourth question are not advantages or disadvantages of the Practical Guide, but of the *software* itself. The ease, clarity and illustrated format were the factors that the participants liked most about the material. Meanwhile, the size of the images and letters were negative points of the Practical Guide, according to some participants. It was also pointed out that the material should have had more content, however, as it was a practical guide, some of *ChemSketch*'s functionalities were not covered.

Question five of the final questionnaire sought to evaluate the mini-course. The question was as follows:

5. How would you rate the Minicourse ChemSketch: Learning to draw organic structures?

(   )Good

(   )Good

(   )Regular

(   )Bad

(   )Bad

Graph 6 shows the results of this question and indicates that the event was rated very positively by the participants.

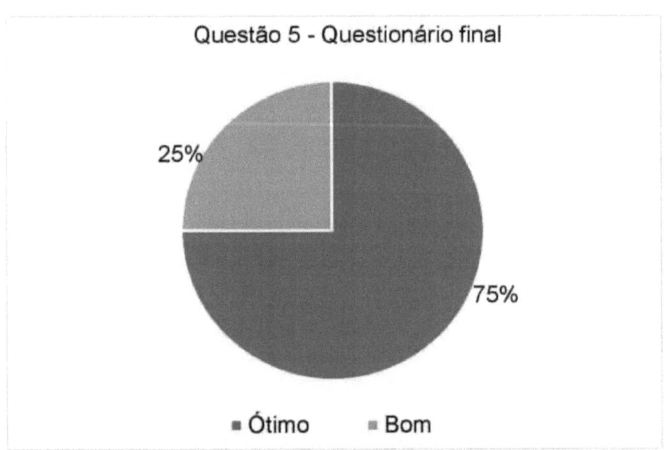

Graph 6. Results of Question 5 - Final questionnaire (G1, G2 and G3). Source: Prepared by the author.

On question six below, all the teachers said that they intend to use *ChemSketch* in their classes. This shows that simple initiatives, such as a mini-course or workshop and the distribution of practical materials, can encourage primary school teachers to use new educational technologies more and in a more productive way.

6. Do you intend to use the ChemSketch software more often in your classes?

( ) Yes   ( ) No   ( ) I don't know

This unanimity indicates that the teachers really do believe that *ChemSketch* will help their practice and their students' learning. When carrying out a study evaluating the use of the *software* with students for the isomerism content, Raupp et al. (2010) stated that:

The use of this *software* apparently helps to internalize these representations, after the student has used them externally, by internalizing the representations and operative invariants associated with the construction of 3D models and the rotation of these models. Eventually, this leads to accommodation, where the pre-existing invariants and representations in the student's cognitive structure are transformed according to the new logic presented by the *software*. (RAUPP et al., 2010, p. 31).

Finally, the last question of the final survey asked participants to rate their mastery of *ChemSketch*.

7. How would you rate your level of mastery of ChemSketch after the mini-course?

( ) Very

( ) Fair

( ) Not much

( ) None

Graph 7 shows that the participants acquired a basic mastery of *ChemSketch*'s functionalities. However, the fact that only 3 out of the 12 participants considered that they were very proficient in the program suggests that the mini-course time of only eight hours was not enough for everyone to effectively master the use of the tool.

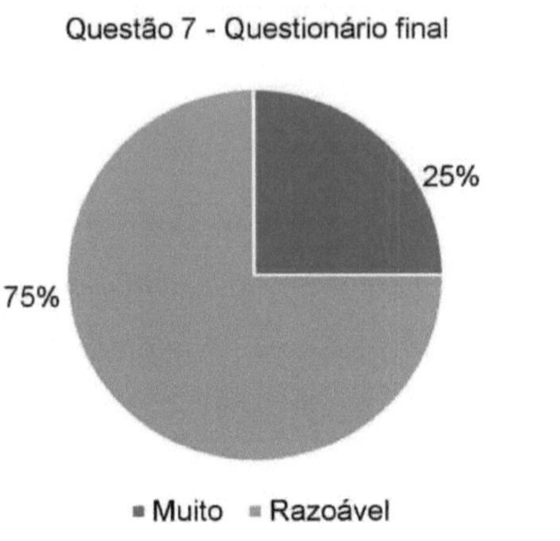

Graph 6. Results of Question 7 - Final questionnaire (G1, G2 and G3). Source: Prepared by the author.

What is worrying about this result is that, because they only consider their mastery to be reasonable, most teachers may still be resistant or unsure about using the *software* in classroom teaching situations. In this way, it is understood that further training with more time spent using the *software* could increase teachers' confidence in using the program with their students.

In this sense, Santiago (2010) says that "overcoming the current teaching model, with its limitations, is one of the most urgent needs for change in the initial and continuing training of chemistry teachers". The mere holding of a mini-course or the distribution of support material will not be enough for all the participants in this work to adopt *ChemSketch* in their teaching practices.

For this to happen, it is extremely important that the teacher himself wants to learn more about the program and that the school offers the conditions and support for digital educational practices to be carried out successfully. Pereira and colleagues (2014) argue that:

Teachers should be very careful when using any technological resource in their teaching methodology, as there are still many students with real difficulties or little familiarity with computers. In addition, they need to constantly adapt/refresh their lesson plans so that they contribute to the assimilation of the content studied. ICTs used in an educational role become truly significant in the teaching and learning process when their functions are in line with their conscious practice and the prior knowledge of the content they are part of has already been acquired in the classroom (PEREIRA et al., 2014, p. 9).

This concern suggests that the use of *ChemSketch*, or any other technological resource, can be positive or negative for student learning. However, a teacher who is aware of and well prepared to use these ICTs can do a lot to ensure that activities in the context of digital education are productive for the student to learn and, consequently, for the teacher's own work.

4.4 Presentation of the proposed activities

Below are the seven activities suggested by the mini-course participants. These activities aim to use *ChemSketch* as a tool in the teaching and learning process of chemistry, and make up the last chapter of the Practical Guide to Using *ChemSketch*.

Suggested activity 1: Continuing education

Carry out ongoing training in schools for teachers in the field of natural sciences, distributing the Practical Guide to Using ChemSketch. It was reported that teachers find it difficult to adopt new technologies and that the inclusion of ChemSketch in ongoing training (horizontal planning) in the school itself would enable teachers to get to know this tool and use it to prepare lessons, demonstrate experiments and in the classroom with students.

Although this activity is not aimed directly at students, it shows the importance of introducing tools such as *ChemSketch* to primary school teachers. For it to be carried out, this action would depend on the willingness of school management and teachers who have already mastered the *software*.

In this case, the Practical Guide could contribute to the teachers' learning in the proposed continuing training, and could serve as support material provided by the school or by the State Department of Education and Sport itself. However, it is important to note that this proposed suggestion is not a practice that uses *ChemSketch* directly, but an initiative that could be adopted by schools to disseminate digital tools like this.

It is possible that some teachers will resist taking part in the suggested continuing training, since not all teachers are able to handle *software*, especially *software* in other languages. However, this could be mitigated by presenting the program's potential and distributing the Practical Guides, which are in Portuguese.

Suggested activity 2: Structures from data

Carry out an activity using four dice: one equivalent to the number of carbons (from 1 to 6), another with bonds (single, double or triple), another with six functional groups and the last with the number of the position of the unsaturations or functional groups on each carbon. The students would throw the dice and, according to the top four faces, they would assemble the respective organic structures on ChemSketch, projecting the construction of the molecules via a Datashow. This activity could be done in groups.

The second activity proposed brings ICT and play closer together. The activity of assembling structures from the dice and their respective correspondences could be interesting and productive for the students. However, the idea of assembling the structures in *ChemSketch*, which are "formed" as the dice are rolled, could attract the students' attention even more. It would also encourage them to master the *software* and generate interest in discovering other functions of the program.

For example, if the results of the dice rolls were:

1st Data: 5 (compound with five carbons)

2nd Data: - (simple bonds)

3rd Data: -OH (hydroxyl)

Data 4: 2 (hydroxyl attached to carbon 2 of the chain)

These results would lead the students to draw a structure in the program that matches the faces of the data. In this case, the compound to be drawn in *ChemSketch* by the students would be pentan-2-ol (Figure 11). If a double or triple bond was drawn on the second dice, dice 4 should be rolled to determine the position of the compound.

If the students were unable to draw the corresponding structures, the teacher could show them what the structure would look like on the board and then draw it in the *software*. The number of throws and dice could further increase the possibilities of this fun activity. The adoption of scores for each success could stimulate the students even more, guaranteeing their involvement and the success of the activity.

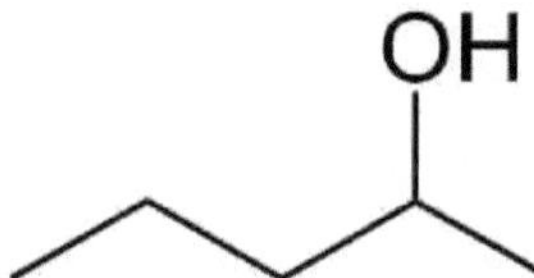

Figure 11. Structure of pentan-2-ol. Source: Prepared by the author.

Suggested activity 3: Making thermochemical graphs

Suggest assembling graphs related to the thermochemistry content. Since ChemSketch makes it possible to draw graphical diagrams, students could create graphical representations of endothermic and exothermic reactions, for example.

ChemSketch's main focus is on drawing organic structures and determining their properties and shapes. As such, its graphical drawing resources are limited. However, as it has an intuitive and relatively simple interface, teachers can guide students to create a variety of qualitative graphs.

For example, the teacher could use *ChemSketch* with the students to explore the graphical information of an exothermic reaction, as shown in Figure 12. In this case, it is possible to visualize the reduction in energy of the products compared to the reactants, as well as graphically showing the activation energy of a generic chemical reaction. This activity could also be used to study gas transformations and even mathematical functions.

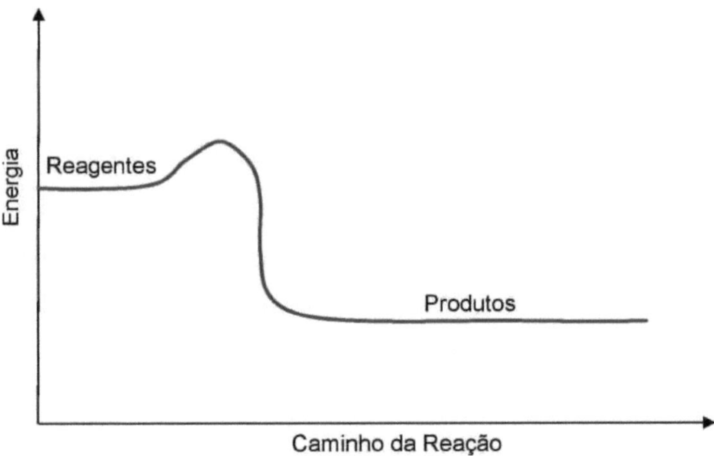

Figure 12. Graph of an exothermic reaction.

Source: Prepared by the author.

Suggested activity 4: Getting to know the laboratory

In the absence of a laboratory and even as a way of preparing students for experimental practices, ChemSketch can be used for activities in which students put together some of the important diagrams present in chemistry laboratories. Students could learn the names of the main glassware and equipment, as well as demonstrating systems such as distillation, filtration, titration, etc. One participant suggested that the students be given mixtures and that they set up, in ChemSketch, a system of glassware and equipment that would allow them to separate each of its components.

ChemSketch has a database with lots of pictures of glassware and other materials commonly used in laboratories (Figure 13). The proposed activity would be a way to prepare students for laboratory classes or to introduce them to the main materials, their names *and* functions.

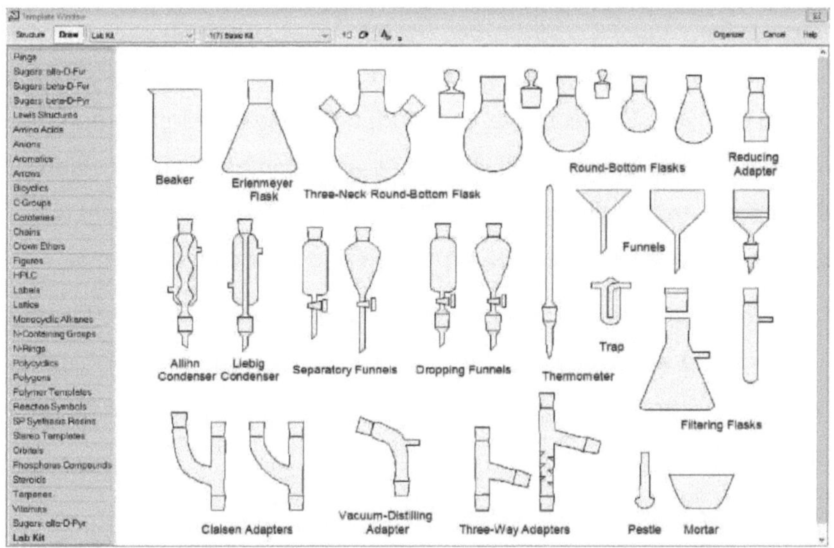

Figure 13. Laboratory glassware kit.

Source: *ChemSketch print screen* on the Windows 10 operating system.

In addition, the activity also proposes that students set up laboratory schemes with the resources available in *ChemSketch.* Simple distillation, filtration and titration systems are some of the possibilities to be created by the students in the program's interface.

Activity suggestion 5: Chemical applications

It was proposed that ChemSketch could be used as the basis for an activity in which students would create apps (for portable devices) that would perform some of the program's functions, such as: the main functional groups, nomenclature rules, calculating the properties of organic compounds, etc. The students would present the apps to the class, and an election could be held to choose the best one.

Developing *software*, such as *ChemSketch*, is a complex process that requires programming professionals and a detailed scientific background. However, the creation of simple applications, especially those used on *smartphones*, are simpler and can be used in teaching and learning situations.

71

The proposed activity makes it clear that *ChemSketch* would only be a basis for the development of the *apps* and, therefore, only some of the program's functions could be used. For example, the students could develop an app that allows them to calculate the molecular masses and elemental composition of substances from the molecular formulas. Or one that shows the different types of organic functions, with examples of structures and their official names.

It's important to make it clear that even without using *ChemSketch*, the students could create apps to present to the class. There are *online* tools that make it easier to create simple *apps*, such as the App Factory tool (http://fabricadeaplicativos.com.br/).

Suggested activity 6: Studying isomerism

Use ChemSketch to approach the subject of isomerism. In this activity, the molecules would be shown on the two-dimensional plane and in three dimensions, showing students the spatial differences that organic compounds with the same molecular formula have. The types of isomerism, especially stereoisomerism, could be presented to the students, allowing them to create and manipulate the structures.

One of the most interesting features of *ChemSketch* is the possibility of manipulating structures in three dimensions. This allows you to work with students on various subjects, such as isomerism. Isomerism occurs when different substances have the same molecular formulae. In constitutional isomerism, the difference between the isomers is in their planar structures. In stereoisomerism, the difference occurs in the way the atoms are arranged in space.

The possibility of visualizing and moving molecular structures can make it easier for students to understand isomerism. In addition, by generating names for the structures, *ChemSketch* identifies whether the isomers are E or Z (in the case of geometric isomerism) or R or S (in the case of optical isomerism). For example, the teacher could suggest discussing the case of thalidomide.

This substance can have sedative or teratogenic effects, depending on the spatial arrangement of its atoms. With *ChemSketch*, the two thalidomides (Figure 14), R and S, could be drawn and visualized three-dimensionally and their differences discussed.

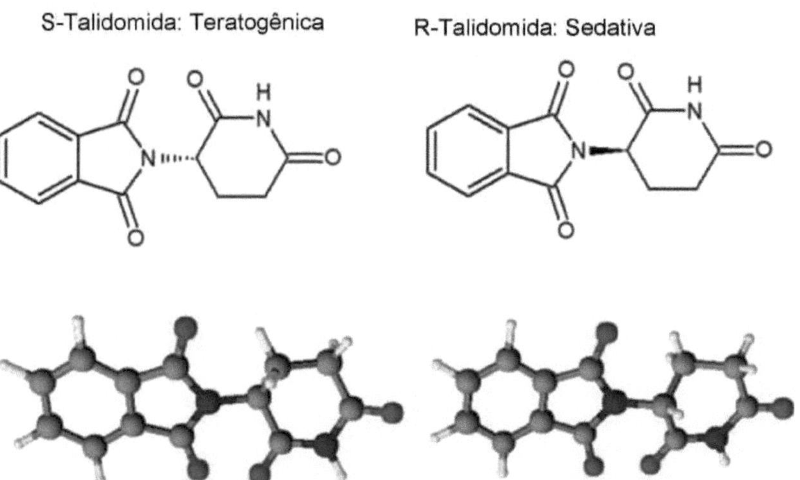

Figure 14. Thalidomide structures. Source: Prepared by the author.

Suggested activity 7: Investigating organic compounds

The last activity suggested was to investigate organic compounds and their properties. The students would form groups and be given an unidentified structure to draw on ChemSketch, with the aim of finding out which substance has that structure and what its main properties and applications are, using the software's chemical database search tools. After a certain amount of time, the groups would present the results of their research to the whole class. It was pointed out that this activity could be used as a way of introducing certain content, such as drugs and vitamins, for example.

This activity would take advantage of one of *ChemSketch*'s most interesting tools, the search for information on the internet about the structures drawn in the program. *ChemSketch* has web access to three databases of chemical structures: *PubChem*, *eMolecules* and *ChemSpider*. The teacher can preselect structures on a topic (drugs, for example) and the students would do their research using only the structures provided by the teacher.

If the structure suggested to one of the groups was the one shown in Figure 15, the students, after drawing the chain, would use the program's three databases (*PubChem*, *eMolecules* and *ChemSpider*) to find out which substance has that structure. With access to the internet, the students would find that it was the structure of ecstasy. With this information, further research could be carried out and each group would present the characteristics and effects of each of the substances suggested by the teacher.

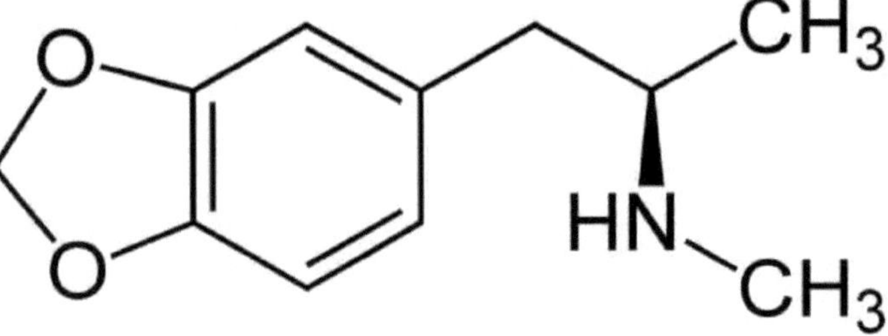

Figure 15. Structure of ecstasy. Source: Prepared by the author.

5 FINAL CONSIDERATIONS

Attributing the success or otherwise of an educational practice to the use of a particular teaching methodology can lead to a misunderstanding of the teaching and learning process. In this sense, experimentation and laboratory practices have already been seen as a kind of solution to get students more interested in the natural sciences, facilitating their learning. More recently, the importance of playful activities has come to the attention of science teaching researchers, and is considered an efficient alternative to traditional didactic practices.

In this respect, both experimentation and playfulness are methodologies that seek to promote students' interest and participation in the construction of their learning. However, the effectiveness of these teaching methods will depend on a series of factors, in which the figure of the teacher is still essential. Accepting also that not every teaching process will culminate in learning, it is understood that any innovative and differentiated educational practices need to be based on clear and pedagogically important foundations.

With this in mind, we understand that the use of information and communication technologies in educational practices is not a magic formula that will make all students want to learn what the teacher wants to teach. However, just like experimentation and play, educational technologies themselves can contribute to the success of the teaching and learning process. For this to happen, it is essential that all those involved in this process, especially the students and teachers, want and are dedicated enough to achieve the goals of education.

This work sought to understand how molecular structuring *software* could be used as a tool in the teaching and learning process of chemistry in Acre. Finding this "how" with infallible precision is an impossible task. It would be presumptuous of any researcher to prescribe ways of teaching that would always guarantee excellent results, regardless of the different contexts that surround a classroom. However, as a chemistry teacher and a citizen concerned about education, it would be inconsistent not to seek to disseminate tools and methodologies that can help other teachers in the arduous task of teaching individuals who sometimes don't even want to learn.

One of these tools is *ADC/Labs ChemSketch freeware*. The preparation of a Practical Guide and a mini-course enabled a small number of teachers and interested parties to get to know the main features of the *software* and to assess whether and how it can be used in teaching and learning situations. The proposals suggested by the teachers themselves show that it is possible to combine *ChemSketch* with the teaching of various subjects, especially organic chemistry.

It is not possible at the moment to assess whether these proposals will be successful when put into practice. However, it was possible to see from the contributions made by the participants in this research that *ChemSketch* is an excellent tool that can make teaching chemistry more attractive and participatory, considering the potential of new educational technologies today.

6 EDUCATIONAL PRODUCT

Title: **Practical Guide to Using ChemSketch**

Descriptive Synopsis: The **Practical Guide to Using ChemSketch** consists of a manual that presents the main features of the program. The Guide covers the process of obtaining and installing *ChemSketch Freeware*, details the function of various features and suggests activities that can be carried out using the *software* in an educational context.

Student author: Alcides Loureiro Santos

Teaching author: Anelise Maria Regiani

Target audience: Primary school chemistry teachers, high school and college students, and anyone else interested in using *ChemSketch Freeware*.

Product URL: **https://goo.gl/tYvavJ**

REFERENCES

ACRE. State Government. State Department of Education. **Curriculum Guidelines for Secondary Education: Notebook 1 - Chemistry.** Rio Branco: SEE, 2010. 44 p.

ANDERY, M. A. **To understand science: a historical perspective.** 6th ed. rev. and expanded. - Rio de Janeiro: Espaço e Tempo: Sâo Paulo: EDUC, 1996.

AZEVEDO, B. F. T. Tópicos em Construçâo de Software Educacional. **Master's Degree in Informatics: Works in Informatics in Education, UFES**, Vitória, July 1997. Available at: <http://www.inf.ufes.br/~tavares/trab3.html>. Accessed on: October 22, 2013.

BENITE, A. M. C.; BENITE, C. R. M. THE COMPUTER IN CHEMISTRY TEACHING: Impressions versus Reality. A Focus on Public Schools in the Baixada Fluminense. **Ensaio Pesquisa em Educaçâo em Ciências**, Vol. 10, No 2 (2008). Available at <http://www.portal.fae.ufmg.br/seer/index.php/ensaio/article/viewArticle/153>. Accessed on March 31, 2015.

BONA, B. O. Analysis of educational software for teaching mathematics in the early years of elementary school. **Experiências em Ensino de Ciências**, vol.4, n. 1, pp.35-55, 2009.

BORGES, J. C. F. **Formação de professores na área de Ciências da Natureza - anâlise de uma pràtica reflexiva no Estâgio Supervisado**.

2010. 175 f. Thesis (Doctorate in Science Education) - Faculty of Sciences, São Paulo State University, Bauru, 2010. Available at: < http://www2.fc.unesp.br/BibliotecaVirtual/DetalhaDocumentoAction.do?idDocumento=378# >. Accessed on August 1, 2014.

BORGMANN, A. **Technology and the Character of Contemporary Life: A Philosophical Inquiry**. The University Chicago Press, 1984. 307p.

BRAZIL. Ministry of Education (MEC). **Curriculum Guidelines for Secondary Education. Natural Sciences, Mathematics and their Technologies** Brasilia: MEC, 2006.

BRAZIL. Ministry of Education (MEC). **Educational Guidelines Complementary to the National Curriculum Parameters (PCN+).** Brasilia: MEC/Semtec, 2002.

BRAZIL. Ministry of Education. Secretariat for Secondary and

Technology. **National curriculum parameters: secondary education.** Brasilia: MEC/SEMTEC, 1999.

CABRERA, W. B.; SALVI, R. Playfulness in High School: Research Aspirations from a Constructivist Perspective. In: ENCONTRO NACIONAL DE PESQUISA EM EDUCAÇAO EM CIENCIAS, 5, 2005, Ijul. **Proceedings.** ijui: Unijui, 2005. 2-4.

CAHAPUZ, A.; PRAIA. J.; JORGE, M. **Da Educaçâo em Ciência às Oriorientações para o Ensino das Ciências: An Epistemological Rethink.** 2004. Available at <http://www.scielo.br/pdf/ciedu/v10n3/05>. Accessed on March 31, 2015.

CARVALHO, A. M. P. Piaget and the teaching of science. **Revista da Faculdade de Educaçâo.** Sâo Paulo: Faculty of Education, uSP, v. 9, p. 55-77, 1983

CHAKUR, C. R. de S. L. **Fundamentos da Pràtica Docente: Por uma Pedagogia Ativa.** Paidéia, FFCLRP, Ribeirâo Preto, 1995.

CHALMERS, A. F. **O que è Ciência afinai?** - 1st ed. Sâo Paulo: Editora Brasiliense, 1993.

CHASSOT, A. Scientific literacy: a possibility for social inclusion. **Revista Brasileira de Educaçâo**, Jan/Feb/Mar/Apr 2003 N° 22. Available at <http://www.scielo.br/pdf/rbedu/n22/n22a09>. Accessed on March 31, 2015.

CuNHA, M. B. **High school students' perception of science and technology and science** communication. Thesis (Doctorate in Science Education) - University of Sâo Paulo, Sâo Paulo, 2009. Available at: < http://www.teses.usp.br/teses/disponiveis/48/48134/tde-02032010-091909/publico/Marcia_Borin_Cunha.pdf>. Accessed on May 29, 2015.

CUPANI, A. Technology as a philosophical problem: three approaches. **Scientiae Studia**. Sâo Paulo, v. 2, n 4, 2004.

DEL PINO, J. C.; FRISON, M. D. CHEMISTRY: A SCIENTIFIC KNOWLEDGE FOR THE FORMATION OF THE CITIZEN. **Revista de Educaçâo, Ciências e Matemàtica.** v.1 n.1 aug/dez. 2011. Available at <http://publicacoes.unigranrio.edu.br/index.php/recm/article/view/1585>. Accessed on March 31, 2015.

EICHLER, M.; PINO, J. C. Del. Computers in Chemistry Education: Atomic Structure and Periodic Table. **Quimica Nova**, Sâo Paulo, v. 23, n. 6, p. 835-9, 2000.

FEENBERG, A. Critical Theory of Technology: An Overview. **TAILORING BIOTECHNOLOGIES**. Vol. I, Issue I, winter 2005, p. 47-64.

FERREIRA, V. F. Interactive technologies in teaching. **Quimica Nova**. 21, 780, 1998.

FRANCELIN, M. M. The epistemology of complexity and information science. **Ci. Inf., Brasilia**, v. 32, n. 2, p. 64-68, May/Aug. 2003.

Available at <http://revista.ibict.br/index.php/ciinf/article/view/118/99>. Accessed on March 31, 2015.

FREITAS, M. T. de A. **Vygotsky and Bakhtin. Psychology and Education: an intertext**. Sâo Paulo: Atica, 1999.

GABINI, W. S. **Continuing education for chemistry teachers: collectively facing the challenge of IT at school**. 2008. 299 f. Thesis (Doctorate in Science Education) - Faculty of Sciences, São Paulo State University, Bauru,

2008. Available at:
<http://www.educadores.diaadia.pr.gov.br//arquivos/File/2010/artigos_teses/2011/quimica/teses/form_cont_prof_quim_tese.pdf>. Accessed on August 7, 2014.

GABINI, W. S. **Informatics and chemistry teaching: investigating the experience of a group of teachers**. 2005. 150 f. Dissertation (Master's Degree in Education for Science) - Faculty of Sciences, São Paulo State University, Bauru, 2005. Available at:

<http://www2.fc.unesp.br/BibliotecaVirtual/ArquivosPDF/DIS_MEST/DIS_MEST20050215_GABINI%20WANDERLEI%20SEBASTIAO.pdf>. Accessed on 07 Aug. 2014.

GEHLEN, S. T.; AUTH, M. A. Contextualization and Meaning in the Teaching of Natural Sciences. In: ENCONTRO NACIONAL DE PESQUISA EM

EDUCATION IN SCIENCES, 5, 2005, Ijui. **Proceedings.** Ijui: Editora Unijui, 2005. 211.

GIORDAN, M.; **Computers and languages in science classes**. Ed. Unijui, 2008

GIORDAN, M. O COMPUTADOR NA EDUCAÇÃO EM CIÊNCIAS: BREVE REVISÃO CRÍTICA ACERCA DE ALGUMAS FORMAS DE UTILIZAÇÃO.

Ciência & Educaçâo, v. 11, n. 2, p. 279-304, 2005.

GIORDAN, M. O papel da Experimentaçâo e Ensino de Ciências. **Quimica Nova na Escola**, Sâo Paulo, n.10, p. 43-9, nov. 1999.

KENSKI, V. M. **Educaçâo e Tecnologias: o novo ritmo da informaçâo.**

Campinas, SP: Papirus, 2007

KOHL, M. O. **Learning and Development**. 1st ed. Sâo Paulo, Scipione, 2000.

LEITE, B. S. **Tecnologias no Ensino de Quimica - Prática e teoria na formaçâo docente**. 1st ed. Curitiba, Appris, 2015.

LEITE, W. S. S.; RIBEIRO, N. C. A. do (2012). The inclusion of ICTs in Brazilian education: problems and challenges. Magis, **Revista Internacional de Investigación em Educación**, 5 (10), 173-187. Available at <http://www.redalyc.org/articulo.oa?id=281024896010>. Accessed on March 31, 2015.

LI, Z.; WAN, H.; SHI, Y.; OUYANG, P. Personal Experience with Four Kinds of Chemical Structure Drawing Software: Review on ChemDraw, ChemWindow, ISIS/Draw, and ChemSketch. **J. Chem. Inf. Comput. Sci**, v. 44, n. 5, p. 1886-1890, 2004.

LOPES, A. O. **Planning teaching from an educational perspective.** In: VEIGA, I. P. A. **Rethinking didactics,** 16ª . Ed. Campinas: Papirus, 2000.

LOPES, J. High School: Brazil's biggest education problem. **Isto É**, year 37, n. 2289, p. 52-56, October 2, 2013.

LÜDKE, M.; ANDRÉ, M. E. D. A. **Pesquisa em educaçâo: abordagens qualitativas**. Sâo Paulo: EPU, 1986.

MAIA, J. de O.; JUNQUEIRA, M. M.; WARTHA, E. J.; SILVA, E. L. **PIAGET, AUSUBEL, VYGOTSKY AND EXPERIMENTATION IN CHEMISTRY TEACHING**. IX CONGRESO INTERNACIONAL SOBRE INVESTIGACIÓN EN DIDACTICA DE LAS CIENCIAS. 2013. Available at <http://congres.manners.es/congres_ciencia/gestio/creacioCD/cd/articulos/art _859.pdf>. Accessed on March 31, 2015.

MALDANER, Otavio Aloisio. **The initial and continuing training of chemistry teachers**. Ijui: UNIJUÌ, 2000.

MARQUES, P. S.; GONÇALVES, I. C. B.; AGUIAR, L. C. da C. Scientific Literacy and Local Knowledge: The Case of Vila do Abraâo, Ilha Grande- RJ. **ACTS AND RESEARCH IN EDUCATION - PPGE/ME FURB**. ISSN 1809-0354 v. 6, n. 2, p. 521-534, mai./ago. 2011. Available at <http://proxy.furb.br/ojs/index.php/atosdepesquisa/article/view/2264/1698>.

Accessed on March 31, 2015.

MATURANA, H. R.; VARELA, F. G. **The Tree of Knowledge**: The biological bases of human understanding. Workshopsy, Campinas, Brazil, 1995.

MELo, E. S. N.; MELo, J. R. F. Simulation software in chemistry teaching: a social representation of teaching practice. **ETD - Educaçâo Temàtica Digital**, Campinas, v.6, n.2, p.51-63, jun. 2005 - ISSN: 1676-2592.

MoRIN, E. **Science with a conscience**. Translated by Maria D. Alexandre and Maria Alice Sampaio Dória. - Revised and modified by the author. - Rio de Janeiro: Bertrand Brasil, 2005. 336p.

MORTIMER, E. F.; MACHADO, A. H. **Quimica**. 1ed. Sâo Paulo: Scipione, 2007. (Teacher's Book).

NARDI, R. (Org.) **Current issues in science teaching.** 2 ed. Sâo Paulo: Escrituras, 2009.

OLIVEIRA, L. B. de. **Action research on practice as an element of initial teacher training through the marine chemistry mini-course**.

2010, 69f. Monografia (Quimica - Licenciatura Plena), Instituto Luterano de Ensino Superior ILES/ULBRA, Itumbiara, 2010.

PAVLINIC, S.; BUCKLEY, P.; BURNS, J. and T. WRIGHT. Computing in stereochemistry - 2D Or 3D representations? In **Research in Science Education - Past, Present, and Future,** Amsterdam: Springer Netherlands, 2007.

PEREIRA, D. I. dos S.; LIMA, B. A. T.; SILVA, T. R.; SANTOS, M. B. H.;

ALMEIDA, R. V. INFORMATION AND COMMUNICATION TECHNOLOGIES IN CHEMISTRY TEACHING. **INTERNATIONAL CONGRESS ON EDUCATION AND INCLUSION**. 2014. Available at <

http://editorarealize.com.br/revistas/cintedi/trabalhos/Modalidade_1datahora_ 08_11_2014_23_52_06_idinscrito_1267_aecb5f140c86ddd5ae7dde49ec86e

381.pdf>. Accessed on 17 Dec. 2015.

PEREIRA, J. R.; ARAÙJO, M. C. P. de. CONCEPTIONS OF SCIENCE: AN EPISTEMOLOGICAL REFLECTION. **VIDYA**, v. 29, n. 2, p. 57-70, jul./dez., 2009 - Santa Maria, 2010. ISSN 2176-4603 X.

PIAGET, J. **Jan Amos Comênio**; translation: Martha Aparecida Santana Marcondes. Pedro Marcondes, Gino Marzio Ciriello Mazzetto; organization: Martha Aparecida Santana Marcondes. - Recife: Joaquim Nabuco Foundation, Massangana Publishing House, 2010.

PIAGET, J. **The birth of intelligence in the child**. 4th ed. Rio de Janeiro, LTC, 1998.

PRIGOGINE, I. & STENGERS, I. **The new alliance: Metamorphosis of science.** Translated by Miguel Faria and Maria Joaquina Machado Trincheira. Review: Joâo Pedro Mendes. Third edition. UnB, 2005.

RAMOS, E. M. F. (Org.). **Informàtica na escola: um olhar multidisciplinar**. Fortaleza: UFC Publishing House, 2003.

RAUPP, D.; SERRANO, A.; MARTINS, T. L. C.; SOUZA, B. C. Using software to build molecular models to teach geometric isomerism: a case study based on cognitive mediation theory.

Revista Electrónica de Ensenanza de las Ciencias, v.9, n.1, p.18-34, 2010.

REZZADORI, C. B. Dal B.; CUNHA, M. B. da. Production of Teaching Materials: A Proposal for Environmental Chemistry. **Varia Scientia**, [s.l.], v. 5, n. 9, p. 177-88, 2005.

SAMPAIO, F. F. (Org.); ELIA, M. F. (Org.). **Projeto Um Computador por Aluno: pesquisas e perspectivas**. 01. ed. Rio de Janeiro: Instituto Tercio Pacitti (NCE) - UFRJ, 2012.

SANTIAGO, E. C. A. **The integration of information and communication technologies in the teaching-learning process in chemistry in public**

schools in Manaus. 2010. 118 f. Dissertation (Professional Master's Degree in Science Teaching in Amazonia) - Escola Normal Superior, Universidade Estadual do Amazonas, Manaus, 2010. Available at: <http://www.pos.uea.edu.br/data/area/titulado/download/16-4.pdf>. Accessed on August 7, 2014.

SANTOS, A. M. P. **Ensino a Distância para Professores - Um Caso Real de Sucesso no âmbito do Programa Prof2000**. 2004. Available at <http://www.abed.org.br/site/pt/midiateca/textos_ead/659/ensino_a_distancia _for_teachers_- _a_real_case_of_success_within_the_program_prof2000_>. Accessed on Aug. 12, 2014.

SANTOS, M. R. C. dos.; AZEVEDO, R. O. M. **INFORMATION AND COMMUNICATION TECHNOLOGIES (ICT) IN THE TEACHING OF CHEMISTRY.** III Encontro Nacional de Ensino de Ciências da Saùde e do Ambiente Niterói/RJ, 2012. Available at <http://ivenecienciassubmissao.uff.br/index.php/ivenecienciassubmissao/ene ciencias2012/paper/download/442/312>. Accessed on March 31, 2015.

SCHNETZLER, R. P. A pesquisa em Ensino de Quimica no Brasil: Conquistas e Perspectivas. **Quimica Nova**, Sâo Paulo, v. 25, supl. 1, p. 1424, 2002.

SERRA, G. M. D. **ICT contributions to science teaching and learning: trends and challenges.** 2009. 383 f. Dissertation (Master's in Education) - Faculty of Education, University of São Paulo, São Paulo, 2010. Available at: <http://www.teses.usp.br/teses/disponiveis/48/48134/tde-05012010-142158/publico/Glades_Miquelina_ME.pdf>. Accessed on August 7, 2014.

SILVA, A. A. da. The Construction of Scientific Knowledge in the Teaching of Chemistry. **Revista Thema**, v. 9, n. 2 (2012a). Available at: <http://revistathema.ifsul.edu.br/index.php/thema/article/view/130>. Accessed on March 31, 2015.

SILVA, E. M. R. TIC NA EDUCAÇÂO: ANALISE PRELIMINAR DOS NOVOS

SABERES DA FORMAÇAO DOCENTE NAS UNIVERSIDADES de SERGIPE. **Revista Contrapontos - Eletrônica**, v. 12, n. 1, p. 37-46, jan-abr 2012b.

SOUZA, F. N. de; BEZERRA, A. C. (Org.). **ICT tools at school: Practical applications.** Aveiro (Portugal): University of Aveiro, 2013. Available at: http://www.bubok.pt/livros/6768/Ferramentas-TIC-na-escola-Aplicacoes-praticas. Accessed on 06 Feb. 2014.

TAVARES, R.; SOUZA, R. O. O.; CORREIA, A. O. A STUDY ON ICT AND THE TEACHING OF CHEMISTRY. SIMTEC **Annals** - ISSN: 2318-3403. Aracaju/SE - 25 to 27/09/ 2013. Vol. 1/n. 1/ p. 657-669. Available at: <http://www.revistageintec.net/portal/index.php/revista/article/download/296/3 46>. Accessed on March 31, 2015.

TERUYA, L. C.; MARSON, G. A.; FERREIRA, C. R.; ARROIO, A. VISUALIZATION IN CHEMISTRY TEACHING: POINTS FOR RESEARCH AND DEVELOPMENT OF EDUCATIONAL RESOURCES. **Quimica Nova**, Sâo Paulo, v. 36, n. 4, p. 561-569, 2013.

VYGOTSKY, L. S. **The social formation of the mind**. 6 ed. Sâo Paulo: Martins Fontes, 1998.

WADSWORTH, B. J. **Inteligência e afetividade da criança na teoria de Jean Piaget**. Sâo Paulo: Pioneira, 1995.

APPENDICES

Initial and final questionnaires

FEDERAL UNIVERSITY OF ACRE

DEAN'S OFFICE FOR RESEARCH AND POSTGRADUATE STUDIES

PROFESSIONAL MASTER'S DEGREE IN SCIENCE AND MATHEMATICS TEACHING

Master's student: Alcides Loureiro Santos

Advisor: Prof. Dr. Anelise Maria Regiani

Title of work: THE USE OF THE CHEMSKETCH SOFTWARE AS A TOOL IN THE TEACHING OF ORGANIC CHEMISTRY IN BASIC EDUCATION IN THE STATE OF ACRE

iNiTiAl QUEsTiOn

1. How important do you think the use of Information and Communication Technologies (ICT) is for teaching chemistry today?

(   )   Very important

(   )   Important

(   )   Not very important

(   )   Not important

2. Do you use chemistry *software* in your classes?

(   ) Yes   (   ) No

If so, which ones do you use the most?

____

3. In your classes, which chemistry subjects would you like to use *software* developed to stimulate your students' learning?

4. Have you taken part in any continuing training courses aimed at introducing ICTs for teaching chemistry?

( ) Yes   ( ) No

5. Do you know the *ChemSketch software*?

( ) Yes   ( ) No

6. How would you rate your level of mastery of *ChemSketch?*

(   )Very

(   )Reasonable

(   )Not much

(   )None

7. Can molecular structure programs contribute to your teaching practice in the classroom and help in the process of teaching and learning chemistry?

(   ) Yes   (   ) No   (   ) Depends

Would you like to comment?

FEDERAL UNIVERSITY OF ACRE

PRO-RECTORY OF RESEARCH AND POSTGRADUATE STUDIES

PROFESSIONAL MASTER'S DEGREE IN SCIENCE AND MATHEMATICS TEACHING

Master's student: Alcides Loureiro Santos

Advisor: Prof. Dr. Anelise Maria Regiani

Title of work: THE USE OF THE CHEMSKETCH SOFTWARE AS A TOOL IN THE TEACHING OF ORGANIC CHEMISTRY IN BASIC EDUCATION IN THE STATE OF ACRE

FiNAL QUEsTioN

1. How can the *ACD/ChemSketch Freeware software be* used as a media tool in the process of teaching Organic Chemistry in High School, considering its potential and limitations, in the context of Basic Education in the state of Acre?

2. What rating would you give *ChemSketch*'s user guide?

(   )Good

(   )Good

(   )Regular

(   )Bad

(   )Bad

3. What did you like most about the Practical Guide to using *ChemSketch*?

4. <u>What did you like least about the Practical Guide to using *ChemSketch*?</u>

5. How would you rate the Minicourse ChemSketch: Learning to draw organic structures?

(   )Good

(   )Good

(   )Regular

(   )Bad

(   )Bad

6. Do you intend to use the *ChemSketch software* more often in your classes?

( ) Yes   ( ) No   ( ) I don't know

7. How would you rate your level of mastery *of ChemSketch* after the mini-course?

(   )Very

(   )Reasonable

(   )Not much

(   )None

yes

I want morebooks!

Buy your books fast and straightforward online - at one of world's fastest growing online book stores! Environmentally sound due to Print-on-Demand technologies.

Buy your books online at
www.morebooks.shop

Kaufen Sie Ihre Bücher schnell und unkompliziert online – auf einer der am schnellsten wachsenden Buchhandelsplattformen weltweit! Dank Print-On-Demand umwelt- und ressourcenschonend produziert.

Bücher schneller online kaufen
www.morebooks.shop

Printed by Books on Demand GmbH, Norderstedt / Germany